Factoring and Operations on Algebraic Fractions

Second Edition

Leon J. Ablon   Sherry Blackman   Helen B. Siner

The College of Staten Island, City University of New York

Anthony Giangrasso

La Guardia Community College, City University of New York

The Benjamin/Cummings Publishing Company, Inc.
Menlo Park, California • Reading, Massachusetts
London • Amsterdam • Don Mills, Ontario • Sydney

Sponsoring Editor: Susan A. Newman
Production Editor: Madeleine Dreyfack
Book Designer: Madeleine Dreyfack
Cover Designer: Judith Sager

TO

Fannie and Meyer Ablon
Clara Kunda
Anthony Jensen Giangrasso
Rigel Cummins

Library of Congress Cataloging in Publication Data
Main entry under title:

Factoring and operations on algebraic fractions.
    (Series in mathematics modules; module 4)
    1. Factors (Algebra).   2. Fractions.   3. Algebra.
I. Ablon, Leon J.   II. Series.
QA39.2.S47 no. 4, 1980 [QA161.F3] 512.9s [512.9'23]
ISBN 0-8053-0134-8   80-24968

ABCDEFGHIJK-DO-89876543210

The Benjamin/Cummings Publishing Company, Inc.
2727 Sand Hill Road
Menlo Park, California 94025

Preface

Purpose of the Steps in Mathematics Modules

This book is one of the Steps in Mathematics Modules. The purpose of this series is to demystify mathematics and provide a foundation for the study of college algebra. Throughout our years of teaching we have become sensitive to the difficulties our students encounter in their mathematics courses. These modules incorporate the techniques we have evolved to overcome these problems.

Features

Development of Concepts: There are very few "rules" in these texts. Whenever possible, we encourage students to figure out what to do from the meaning of the symbols.

Writing Style: We have been as clear as possible in presenting each mathematical concept. We have found that students relate well to our use of everyday language. Our guiding principle has been to use language and syntax no more complex than is necessary to convey each concept. It is the kind of language that is used in the classroom, but is rarely found in mathematics textbooks.

Organization: Each module is divided into eight lessons. Each lesson is designed to be covered in one class period.

Examples and Exercises: We have used many "worked out" examples in each lesson to demonstrate the mathematics. Each lesson ends with two sets of *graded* exercises. The first set is followed immediately by handwritten solutions, so the student can see what a typical solution might look like. The second set, or "additional" exercise set, does not have solutions in the module. The answers to these exercises can be found in the Rationale, a supplement to these modules.

How to Use the Modules

Pacing: Each module is made up of eight lessons, and contains material sufficient for a 40 to 50-minute class period. The background of some students may, of course, dictate a slower or faster pace. On the average we have found that each module can be completed in 10 to 12 class meetings, including time for review and examination.

Planning a Syllabus for a Lecture Mode: The chart below may assist you in setting up a program for a term. It shows the number of modules that can be covered in 10, 12 and 14-week terms with classes that meet 3, 4 and 5 times a week.

Weeks per term	Meetings per week		
	3	4	5
10	3 modules	3 or 4 modules	4 or 5 modules
12	3 modules	4 modules	5 modules
14	4 modules	5 modules	6 or 7 modules

Self-Study or Lab Mode of Instruction: The modules can be used in math lab or for independent study. The preceding chart serves as a guide for students to gauge their progress.

Flexibility: The modular presentation of material allows for flexibility in teaching. In particular, the concept of changing the form of expressions (Modules 2 and 4) is separated from the concept of solving equations (Modules 3 and 5).

A chart of the modules interdependence can be found in the Rationale. We recommend that each module be studied from cover to cover. This gives students a feeling of completeness and insures that they are prepared for succeeding modules.

The Diagnostic Test included in the Rationale is designed to assist in determining an appropriate starting module for each student.

Improvements in This Edition

The first edition of these modules has been used successfully by us and many other math teachers in classrooms and labs for over eight years. As a result of our experiences and helpful feedback from other teachers, we have made several changes and refinements in this second edition.

Decimals: This new edition integrates a review of decimals into both the lessons and homework exercises.

Word Problems: Word problems have been carefully integrated throughout the modules in the examples and exercises.

Exercises: We have increased the number of homework exercises by 20%.

Geometry: The series now includes a review of the concepts of perimeter, area, formulas and the Pythagorean theorem.

Design and Appearance: The modules are typeset, a pleasant change from the camera-ready typewriter version of the first edition. The use of italic and boldface type enables the student to quickly pick out examples, important concepts and key terms.

The Steps in Mathematics Modules Series

The first five steps in mathematics are the equivalent of an elementary algebra course: They are:

Module 1—Operations on Numbers

Module 2—Operations on Polynomials

Module 3—Linear Equations and Lines

Module 4—Factoring and Operations on Algebraic Fractions

Module 5—Quadratic Equations and Curves

Program Rationale and Tests—an instructor's supplement which contains a brief description of the content of each lesson, answers to the Additional Exercises, diagnostic placement examinations and three mastery tests for each of the five modules.

Additional modules in the **Steps in Mathematics Series** cover topics in inter-mediate algebra and other specialized topics.

Module 1a—Practical Mathematics

Module 2a—Practical Mathematics

Module 6 —Basic Trigonometry, second edition

Module 7 —Trigonometry with Applications

Module 8 —Exponents and Logarithms

Module 9 —Advanced Algebraic Techniques

Module 10—Functions and Word Problems

Module 11—Graphing Functions

Module M —Medical Dosage Calculations, second edition

Module SI—Metric System

Acknowledgements

We wish to thank all the people who contributed to the revision of Modules 1–5 with special thanks to our reviewers:

Dr. George Bergman, University of California, Berkeley
Dr. Una Bray, Marymount Manhattan College
Professor Ruth Dorsett, Atlanta Junior College
Dr. Susan Lawrence, New York University
Dr. Betty Philips, Michigan State University
Professor Michael Shaughnessy, Oregon State University
Professor Billie J. Stacey, Sinclair Community College

Our appreciation, also, to Susan Newman, sponsoring editor, and Madeleine Dreyfack, production editor and designer. The special efforts they made to translate our ideas into a finished work are greatly appreciated.

Leon J. Ablon
Sherry Blackman
Anthony P. Giangrasso
Helen B. Siner
October, 1980

To the Student

Read this text. It was written with the help of our students so students could read it. Our students tell us that the best way to learn this material is:

1. Read the lesson.
2. Work out each EXAMPLE yourself.
3. Try the first set of EXERCISES at the end of the lesson and compare your answers with ours.
4. Do the ADDITIONAL EXERCISES for more practice.

We wish you success, and we would like to hear your reactions to the text and your suggestions for future editions.

Contents

Lesson 1   **Factoring and Simplifying Algebraic Fractions**   **1**

Exercises   9
Answers to Exercises   11
Additional Exercises   13

Lesson 2   **FOIL Multiplication and Factoring Trinomials**   **15**

Exercises   25
Answers to Exercises   27
Additional Exercises   30

Lesson 3   **More Factoring**   **32**

Exercises   41
Answers to Exercises   42
Additional Exercises   44

Lesson 4   **Still More Factoring**   **45**

Exercises   52
Answers to Exercises   54
Additional Exercises   56

Lesson 5   **You Guessed It! Factoring**   **57**

Fractions with Opposites   60
Exercises   63
Answers to Exercises   65
Additional Exercises   68

Lesson 6   **Multiplication and Division of Algebraic Fractions**   **70**

Multiplication of Algebraic Fractions   70
Division of Algebraic Fractions   74
Exercises   78
Answers to Exercises   81
Additional Exercises   86

Lesson 7   **Addition and Subtraction of Algebraic Fractions   89**

Fractions with Identical Denominators   89
Fractions with Different Denominators   92
Exercises   100
Answers to Exercises   103
Additional Exercises   110

Lesson 8   **Advanced Addition and Subtraction of
Algebraic Fractions   113**

Exercises   122
Answers to Exercises   124
Additional Exercises   132

Index   **134**

Factoring and Simplifying Algebraic Fractions

Factoring means splitting a number or expression into two or more parts, which when <u>multiplied</u> together give back the original number or expression. The parts are called **factors**.

Let's look at the expression $3a + 3b$. We can split $3a + 3b$ into two parts; 3 and $(a + b)$. Multiplying 3 by $(a + b)$ gives the original expression $3a + 3b$. 3 and $(a + b)$ are factors of $3a + 3b$. So $3a + 3b$ can be written as $3(a + b)$.

It is important to know how to go back and forth between these two equivalent forms. Sometimes it's useful to have an expression that is a sum, like $3a + 3b$; and sometimes it's convenient to have an expression that is a product, like $3(a + b)$.

$$
\underbrace{3a + 3b}_{\text{Sum}} \xrightleftharpoons[\text{multiply}]{\text{factor}} \underbrace{3(a + b)}_{\text{Product}}
$$

If we start with $3a + 3b$ and get $3(a + b)$, we are factoring. If we start with $3(a + b)$ and get $3a + 3b$, we are multiplying. In the next two examples we review the factoring we did in Module II.

1

EXAMPLE 1   Factor $2x^3 + 4x^2y$.

$$2x^3 + 4x^2y$$

is the same as $(2)(x)(x)(x) + (2)(2)(x)(x)(y)$

Notice that we have split the first term into four parts (factors): 2, x, x, and x. We have split the second term into five parts (factors): 2, 2, x, x, and y. Since 2, x, and x are factors common to both $2x^3$ and $4x^2y$, and since there are no other factors common to these terms, we call $2xx$ or $2x^2$ **the complete common factor** of $2x^3$ and $4x^2y$. So $2x^3 + 4x^2y$ is factored completely as $2x^2(x + 2y)$.

EXAMPLE 2   Factor $4a^2b - 6ab^2 + 2ab$ completely.

$$4a^2b - 6ab^2 + 2ab$$

can be written as $(2)(2)aab - (2)(3)abb + 2ab$

It is convenient to underline the common factors as follows.

$$(2)\ (2)\ \underline{aab} - (2)\ (3)\ \underline{abb} + \underline{2ab}$$

The only common factors are 2 and a and b. Therefore the complete common factor is $2ab$, and the factored form of $4a^2b - 6ab^2 + 2ab$ is

$$2ab(2a - 3b + 1)$$

Let's check this result by multiplying.

$$2ab(2a - 3b + 1) = 4a^2b - 6ab^2 + 2ab$$

In the rest of this lesson, we will use factoring to help us simplify our work with algebraic fractions. In all of our fractions, the bottom will never stand for zero, since division by zero has no meaning. When we are given an algebraic fraction, factor the top and bottom completely. If the same factor is on top and on bottom, we can cancel.

EXAMPLE 3   Simplify $\dfrac{3b + 6c}{9d}$.

First completely factor the top and completely factor the bottom.

$$\frac{3b + 6c}{9d}$$

is factored as $\dfrac{3(b + 2c)}{(3)(3)d}$ $3(b + 2c\ )$

which is $\dfrac{(3)}{(3)} \cdot \dfrac{(b + 2c)}{(3d)}$

But $\dfrac{(3)}{(3)}$ is $\dfrac{1}{1}$ or 1, so we can replace it by 1. That is, we cancel the 3's.

Our symbol for canceling is the slash, which we use either

this way $\dfrac{\cancel{a}^{1}}{\cancel{a}_{1}}$ or this way $\cancel{\dfrac{a}{a}}\,^{1}_{1}$

$$\frac{\cancel{(3)}^{1}}{\cancel{(3)}_{1}} \cdot \frac{(b + 2c)}{(3d)}$$

is the same as $(1) \cdot \dfrac{(b + 2c)}{(3d)}$ or $\dfrac{b + 2c}{3d}$

So $\dfrac{3b + 6c}{9d} = \dfrac{b + 2c}{3d}$.

EXAMPLE 4   Simplify $\dfrac{2a + 2b}{3a + 3b}$. $\dfrac{2(a+b)}{3(a+b)} = \dfrac{2}{3}$

First completely factor the top, then completely factor the bottom.

$$\frac{2a + 2b}{3a + 3b}$$

is factored as $\dfrac{2(a + b)}{3(a + b)}$

But $\dfrac{(a + b)}{(a + b)}$ is 1, because any number (except 0) divided by itself is 1.

So we can cancel.

$$\frac{2\cancel{(a+b)}^{\;1}}{3\cancel{(a+b)}_{\;1}}$$

is the same as $\frac{2}{3} \cdot 1$ or $\frac{2}{3}$

So $\frac{2a+2b}{3a+3b} = \frac{2}{3}$.

EXAMPLE 5 Simplify $\frac{x^2+xy}{x+y}$. $\frac{x(x+y)}{x+y} = x$

First completely factor top and bottom as follows.

$$\frac{x^2+xy}{x+y}$$

is factored as $\frac{x(x+y)}{x+y}$

Note: The bottom of this fraction cannot be factored. We call $x+y$ **prime**.

$$\frac{x\cancel{(x+y)}^{\;1}}{\cancel{(x+y)}_{\;1}}$$ $x \cdot 1 = x$

is the same as $x(1)$ or x

So $\frac{x^2+xy}{x+y} = x.$

EXAMPLE 6 Simplify $\dfrac{ax^2 + ax - a}{bx^2 + bx - b}$.

This can be factored as

$$\frac{a(x^2 + x - 1)}{b(x^2 + x - 1)}$$

But $\dfrac{(x^2 + x - 1)}{(x^2 + x - 1)}$ is 1, because any number (except 0) divided by

itself is 1. We can cancel the trinomials.

$$\frac{a(\cancel{x^2 + x - 1})}{b(\cancel{x^2 + x - 1})} = \frac{a}{b} \quad (1)$$

So $\dfrac{ax^2 + ax - a}{bx^2 + bx - b} = \dfrac{a}{b}$.

$$\frac{a(x + x - 1)}{b(x + x - 1)}$$

Let's look at $\dfrac{a + 2}{a + 3}$. Since $(a + 2)$ and $(a + 3)$ are prime, they cannot be

factored further. So the only factor on top is $(a + 2)$ and the only factor on the bottom is $(a + 3)$. $(a + 2)$ and $(a + 3)$ are not the same.

So when we divide them we don't get 1. We cannot cancel.

Watch it! In $\dfrac{a + 2}{a + 3}$ we cannot cancel the a's either. $\dfrac{a + 2}{a + 3}$ is not the same

as $\dfrac{2}{3}$. For example, if a stands for 5,

then $\qquad \dfrac{a + 2}{a + 3}$

stands for $\qquad \dfrac{5 + 2}{5 + 3}$

or $\qquad \dfrac{7}{8}$

So $\qquad \dfrac{a + 2}{a + 3}$ cannot be simplified.

EXAMPLE 7 Simplify $\dfrac{3x + 6y}{3x - 6y}$.

This can be factored as $\dfrac{3(x + 2y)}{3(x - 2y)}$, and we can cancel the 3's.

$$\dfrac{\overset{1}{\cancel{3}}}{\underset{1}{\cancel{3}}} \cdot \dfrac{(x + 2y)}{(x - 2y)}$$

is the same as $\quad 1 \cdot \dfrac{(x + 2y)}{(x - 2y)}$

or $\qquad\qquad \dfrac{x + 2y}{x - 2y}$

$(x + 2y)$ and $(x - 2y)$ are not the same. We cannot cancel them.

So $\quad \dfrac{3x + 6y}{3x - 6y} = \dfrac{x + 2y}{x - 2y}$.

EXAMPLE 8 Simplify $\dfrac{(x + 2)(x + 1)}{(x + 2)^2}$.

This is the same as

$$\dfrac{(x + 2)(x + 1)}{(x + 2)(x + 2)}$$

We can cancel.

$$\dfrac{\overset{1}{\cancel{(x + 2)}}(x + 1)}{\underset{1}{\cancel{(x + 2)}}(x + 2)}$$

is the same as $\quad \dfrac{x + 1}{x + 2}$

This is as far as we can go. So $\dfrac{(x + 2)(x + 1)}{(x + 2)^2} = \dfrac{x + 1}{x + 2}$.

EXAMPLE 9   Simplify $\dfrac{2x + 4}{10 + 5x}$.

This can be factored as

$$\dfrac{2(x + 2)}{5(2 + x)}$$

But $\underline{2 + x}$ is the same as $x + 2$, so we can write $\dfrac{2(x + 2)}{5(\underline{2 + x})}$ as $\dfrac{2(x + 2)}{5(\underline{x + 2})}$.

We can cancel.

$$2 + x \;=\; x + 2$$

$$\dfrac{2\cancel{(x + 2)}^{\,1}}{5\cancel{(x + 2)}_{\,1}}$$

is the same as $\dfrac{2}{5} \cdot 1$ or $\dfrac{2}{5}$

So $\dfrac{2x + 4}{10 + 5x} = \dfrac{2}{5}$.

EXAMPLE 10   Simplify $\dfrac{2y - 4xy + 6x^2y}{5x - 10x^2 + 15x^3}$. $2y(1 - 2x + 3x^2)$

This can be factored as

$$\dfrac{2y(1 - 2x + 3x^2)}{5x(1 - 2x + 3x^2)}$$

We can cancel the trinomials.

$$\dfrac{2y(1 - \cancel{2x + 3x^2})^{\,1}}{5x\cancel{(1 - 2x + 3x^2)}_{\,1}}$$

is the same as $\dfrac{2y}{5x}$

So $\dfrac{2y - 4xy + 6x^2y}{5x - 10x^2 + 15x^3} = \dfrac{2y}{5x}$.

EXAMPLE 11 Simplify $\dfrac{6x - 12y}{2x + 4y}$.

$\dfrac{6(x - 2y)}{2(x + 2y)}$

This can be factored as

$$\frac{6(x - 2y)}{2(x + 2y)}$$

$(2)(3)\,(x - 2y)$
$(2)\quad(x + 2y)$

This is the same as

$$\frac{2(3)(x - 2y)}{2(x + 2y)}$$

We can cancel the 2's.

$$\frac{\overset{1}{\cancel{2}}(3)(x - 2y)}{\underset{1}{\cancel{2}}\ (x + 2y)}$$

is the same as $\dfrac{3(x - 2y)}{(x + 2y)}$

But $(x - 2y)$ and $(x + 2y)$ are not the same. We cannot simplify

$\dfrac{3(x - 2y)}{x + 2y}$ any further, so $\dfrac{6x - 12y}{2x + 4y} = \dfrac{3(x - 2y)}{x + 2y}$.

Exercises

Factor completely.

1. $3a^2b^2 + 6ab$
2. $5x^2y - 10xy^2 + 5xy$
3. $9ab + 18a^2b^2 - 9ab^2$
4. $4x^2y^2 - 2xy + 8x^2y$
5. $2x^{10}y^{12} + 4x^{12}y^{17} - 6x^{20}y^{13}$

Simplify each of the following factors.

6. $\dfrac{2x + 10y}{4z}$

7. $\dfrac{3a - 6b}{4a - 8b}$

8. $\dfrac{b^3 + bx}{b^2 + x}$

9. $\dfrac{9x^2 + 9x - 9}{ax^2 + ax - a}$

10. $\dfrac{4y^3 + 4y^2 - 4y}{9y^2 + 9y - 9}$

11. $\dfrac{ax^2 - ax}{ax^2 + ax}$

12. $\dfrac{2x - 5y}{4x^2 - 10xy}$

13. $\dfrac{(x - 2y)(x + 3y)}{(x + 3y)^2}$

14. $\dfrac{2x + 4}{4x + 12}$

15. $\dfrac{8x^2 + 4xy - 2x}{12x^2 - 3x + 6xy}$

16. $\dfrac{3a + 9}{12 + 4a}$

17. $\dfrac{(3a - 2b)(10a - 2b)}{(3a - 2b)(5a - b)}$

18. $\dfrac{x^2 - 2x + 3xy}{x - 2 + 3y}$

19. $\dfrac{bx + 2by}{ax + 2ay}$

20. $\dfrac{(3x + 3y)(x + y)}{x + y}$

Answers to Exercises

① $3\underline{a}a\ \underline{b}b + 2\cdot 3\ \underline{a}\ \underline{b}$

  $3ab\,(ab+2)$

② $5\underline{x}x y - 5\cdot 2\underline{x}yy + 5\underline{x}\underline{y}$

  $5xy\,(x-2y+1)$

③ $9\underline{a}\underline{b} + 9\cdot 2\underline{a}a\ \underline{b}b - 9\underline{a}\underline{b}b$

  $9ab\,(1+2ab-b)$

④ $2\cdot 2\cdot x x yy - 2xy + 2\cdot 2\cdot 2\ x x y$

  $2xy\,(2xy-1+4x)$

⑤ $2\ x^{10}y^{12} + 2\cdot 2\ x^{10}x^{2}y^{12}y^{5} - 2\cdot 3\ x^{10}x^{10}y^{12}y^{1}$

  $2x^{10}y^{12}\,(1+2x^{2}y^{5}-3x^{10}y^{1})$

⑥ $\dfrac{2(x+5y)}{2\cdot 2z} = \dfrac{x+5y}{2z}$ ⑨ $\dfrac{9(x^{2}+x-1)}{a(x^{2}+x-1)} = \dfrac{9}{a}$

⑦ $\dfrac{3(a-2b)}{4(a-2b)} = \dfrac{3}{4}$ ⑩ $\dfrac{4y\,(y^{2}+y-1)}{9\,(y^{2}+y-1)} = \dfrac{4y}{9}$

⑧ $\dfrac{b(b^{2}+x)}{(b^{2}+x)} = \dfrac{b}{1} = b$ ⑪ $\dfrac{a\,x\,(x-1)}{a\,x\,(x+1)} = \dfrac{x-1}{x+1}$

12. $\dfrac{\cancel{(2x-5y)}}{2x\cancel{(2x-5y)}} = \dfrac{1}{2x}$

13. $\dfrac{(x-2y)\cancel{(x+3y)}}{(x+3y)\cancel{(x+3y)}} = \dfrac{x-2y}{x+3y}$

14. $\dfrac{2(x+2)}{4(x+3)} = \dfrac{\cancel{2}(x+2)}{2\cdot\cancel{2}(x+3)} = \dfrac{x+2}{2(x+3)}$

15. $\dfrac{2\cancel{x}\cancel{(4x+2y-1)}}{3\cancel{x}\cancel{(4x-1+2y)}} = \dfrac{2}{3}$

16. $\dfrac{3\cancel{(a+3)}}{4\cancel{(3+a)}} = \dfrac{3}{4}$

17. $\dfrac{\cancel{(3a-2b)}(2)\cancel{(5a-b)}}{\cancel{(3a-2b)}\cancel{(5a-b)}} = \dfrac{2}{1} = 2$

18. $\dfrac{x\cancel{(x-2+3y)}}{\cancel{(x-2+3y)}} = \dfrac{x}{1} = x$

19. $\dfrac{b\cancel{(x+2y)}}{a\cancel{(x+2y)}} = \dfrac{b}{a}$

20. $\dfrac{3\cancel{(x+y)}(x+y)}{\cancel{(x+y)}} = \dfrac{3(x+y)}{1} = 3(x+y)$

Additional Exercises

Factor completely.

1. $5x^2y + 10xy^2$
2. $4ab^2 - 8a^2b + 12ab$
3. $7xy - 14x^3y + 21x^2y^2$
4. $3c^4d^3 - 6c^2d^2 - 3cd$
5. $2x^{15}y^{10}z^8 - 8x^{12}y^{10}z^{12} + 4x^{19}y^{14}z^{10}$

Simplify each of the following fractions.

6. $\dfrac{4x + 8y}{12z}$

7. $\dfrac{5x + 5y}{7x + 7y}$

8. $\dfrac{a^2 + 2ab}{a + 2b}$

9. $\dfrac{x^2 + 3xy}{xy + 3y^2}$

10. $\dfrac{2a^2 - 4a - 6}{a^2b - 2ab - 3b}$

11. $\dfrac{2x^2 + 6xy}{3xy + 9y^2}$

12. $\dfrac{3a - 4b}{6a^2 - 8ab}$

13. $\dfrac{(x - 4y)(x + 5y)}{(x - 4y)^2}$

14. $\dfrac{2x - 2y}{x - y}$

15. $\dfrac{4x + 12}{15 + 5x}$

16. $\dfrac{12 + 3a}{7a + 28}$

17. $\dfrac{(a - b)(4a - 3b)}{(a - b)(4a - 3b)}$

18. $\dfrac{3x^2 + 6x + 9}{x^2 + 2x + 3}$

19. $\dfrac{12a - 6b}{3a - 3b}$

20. $\dfrac{4x^2 - 4xy}{4x^2 - 2xy}$

FOIL Multiplication and Factoring Trinomials

FOIL Multiplication

In this lesson we will use a shortcut for multiplying two binomials. The shortcut is called **FOIL**.

First let's multiply $x + 3$ by $x + 2$ the long way.

$$
\begin{array}{r}
x + 3 \\
x + 2 \\
\hline
x^2 + 3x \\
+\ 2x + 6 \\
\hline
x^2 + 5x + 6
\end{array}
$$

Now let's see where each term of $x^2 + 5x + 6$ came from.

$x + 3$
$x + 2$
$x^2 + 3x$
$ + 2x + 6$
$x^2 + 5x + 6$

FIRST

$(x + 3)(x + 2)$

First Terms

$x + 3$
$x + 2$
$x^2 + 3x$
$ + 2x + 6$
$x^2 + 5x + 6$

OUTER

$(x + 3)(x + 2)$

Outer Terms

$x + 3$
$x + 2$
$x^2 + 3x$
$ + 2x + 6$
$x^2 + 5x + 6$

INNER

$(x + 3)(x + 2)$

Inner Terms

$x + 3$
$x + 2$
$x^2 + 3x$
$ + 2x + 6$
$x^2 + 5x + 6$

LAST

$(x + 3)(x + 2)$

Last Terms

FOIL will stand for the following.

First → multiply the two first terms
Outer → multiply the two outer terms
Inner → multiply the two inner terms
Last → multiply the two last terms

Now, let's multiply $x + 3$ by $x + 2$ again, using the shortcut FOIL method.

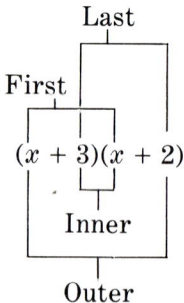

Last

First

$(x + 3)(x + 2)$

Inner

Outer

First + Outer + Inner + Last

x^2 + $2x$ + $3x$ + 6

is the same as $x^2 + 5x + 6$

Note: We added the Outer ($2x$) and the Inner ($3x$) and got $5x$.

EXAMPLE 1 Multiply $x + 4$ by $x + 6$.

Let's do this two ways. First we'll multiply it the long way.

$$
\begin{array}{r}
x + 4 \\
x + 6 \\
\hline
x^2 + 4x \\
+ 6x + 24 \\
\hline
x^2 + 10x + 24
\end{array}
$$

Now, let's use the FOIL method.

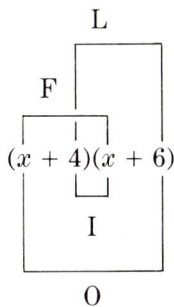

L

F

$(x + 4)(x + 6)$

I

O

F + O + I + L

x^2 + $6x$ + $4x$ + 24

which is $x^2 + 10x + 24.$

EXAMPLE 2 Multiply $x + 5$ by $x - 3$.

We use the FOIL method here.

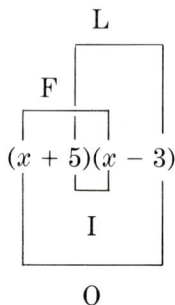

$$(x + 5)(x - 3)$$

$$F + O + I + L$$

$$x^2 - 3x + 5x - 15$$

which is $x^2 + 2x - 15.$

EXAMPLE 3 Multiply $x - \dfrac{1}{4}$ by $x - \dfrac{3}{4}$ using the FOIL method.

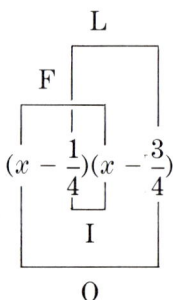

$$\left(x - \tfrac{1}{4}\right)\left(x - \tfrac{3}{4}\right)$$

$$F + O + I + L$$

$$x^2 - \frac{3}{4}x - \frac{1}{4}x + \frac{3}{16}$$

which is $x^2 - x + \dfrac{3}{16}.$

EXAMPLE 4   Multiply $2x - 3$ by $x + 2$ using the FOIL method.

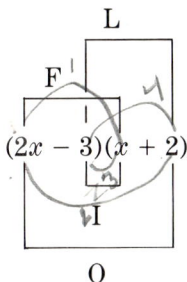

$$(2x - 3)(x + 2)$$

$$2x^2 + 4x - 3x - 6$$
which is $2x^2 + x - 6$.

EXAMPLE 5   Multiply $5x + 6$ by $2x - 3$ using the FOIL method.

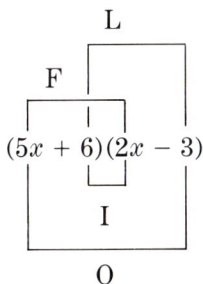

$$(5x + 6)(2x - 3)$$

$$10x^2 - 15x + 12x - 18$$

which is $10x^2 - 3x - 18$.

EXAMPLE 6   Multiply $x + 3$ by $x - 3$ using the FOIL method.

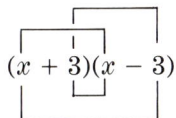

$$(x + 3)(x - 3)$$

$$x^2 - 3x + 3x - 9$$

which is $x^2 - 9$.

Note: $-3x + 3x$ is 0.

EXAMPLE 7   Multiply $2x + 0.5$ by $2x - 0.5$ using the FOIL method.

$$(2x + 0.5)(2x - 0.5)$$

$4x^2 - 1.0x + 1.0x - 0.25$

which is $4x^2 - 0.25$.

EXAMPLE 8   Square $x + 5$ using the FOIL method.

$$(x + 5)^2 = (x + 5)(x + 5)$$

$$= x^2 + 5x + 5x + 25$$

$$= x^2 + 10x + 25$$

EXAMPLE 9   Square $x + y$ using the FOIL method.

$$(x + y)^2 = (x + y)(x + y)$$

$$= x^2 + xy + xy + y^2$$

$$= x^2 + 2xy + y^2$$

EXAMPLE 10   Multiply $3x + 4y$ by $2x - 5y$.

$$(3x + 4y)(2x - 5y)$$

$6x^2 - 15xy + 8xy - 20y^2$

which is $6x^2 - 7xy - 20y^2$.

Factoring

We can undo every multiplication by factoring.

$$(x + 3)(x + 1) \xrightleftharpoons[\text{factor}]{\text{multiply}} x^2 + 4x + 3$$

Now let's start with $x^2 + 4x + 3$ and try to factor it. Whenever you factor, first factor out common monomial factors (as in Lesson 1) if there are any. Then look for binomial factors. In this example, there are no monomial factors, so we should look for something like this:

$$x^2 + 4x + 3 = (?\quad ?)(?\quad ?)$$

Notice that in all the examples of this lesson, the term with the exponent 2 comes from multiplying the first terms using the FOIL method.

$$x^2 + 4x + 3 = (x\quad ?)(x\quad ?)$$

In the same way, the 3 must come from multiplying the last terms using the FOIL method.

$$x^2 + 4x + 3 = (x\quad 3)(x\quad 1)$$

We used the x^2 to get the first terms of the factors (the x's); we used the $+3$ to get the last terms of the factors (the 3 and the 1). Now we use the $+4x$. The $+4x$ comes from the outer and inner terms.

Outer

$$x^2 + 4x + 3 = (x + 3)(x + 1)$$

Inner

The outer terms give us $+x$ since $(x)(+1) = +x$. The inner terms give us $+3x$ since $(+3)(x) = +3x$. These add up to $+4x$ as required.

Now let's use the FOIL method of multiplying to check.

$$(x + 3)(x + 1) = x^2 + x + 3x + 3$$

$$= x^2 + 4x + 3 \qquad \text{It checks!}$$

Let's try another one. Factor $x^2 - 4x + 3$. Since there are no common factors, we look for binomial factors.

$$x^2 - 4x + 3 = (?\quad ?)(?\quad ?)$$

As before the x^2 comes from the first terms.

$$x^2 - 4x + 3 = (x\quad ?)(x\quad ?)$$

And 3 must come from the last terms.

$$x^2 - 4x + 3 = (x\quad 3)(x\quad 1)$$

Now we use the $-4x$, which comes from the inner and outer terms. This time we use negative signs.

$$x^2 - 4x + 3 = (x - 3)(x - 1)$$

The outer terms give us $-x$. The inner terms give us $-3x$. These add up to $-4x$ as required. So $x^2 - 4x + 3$ can be factored as $(x - 3)(x - 1)$. This can also be written as $(x - 1)(x - 3)$.

EXAMPLE 11   Factor $x^2 + 2x + 1$.

There is no common monomial factor, so we go on to the next step and look for binomial factors. The x^2 comes from the first terms.

$x^2 + 2x + 1 = (x \quad ?)(x \quad ?)$

The 1 comes from the last terms.

$x^2 + 2x + 1 = (x \quad 1)(x \quad 1)$

Now we use the $+2x$, which comes from the inner and outer terms.

$x^2 + 2x + 1 = (x + 1)(x + 1)$

So $x^2 + 2x + 1$ can be factored as $(x + 1)(x + 1)$ or $(x + 1)^2$.

EXAMPLE 12   Factor $2x^2 - 12x + 10$.

This time there is a common monomial factor, 2.

$2x^2 - 12x + 10 = 2(x^2 - 6x + 5)$

Now let's factor what's inside the parentheses. The x^2 comes from the first terms.

$2(x^2 - 6x + 5) = 2(x \quad ?)(x \quad ?)$

The 5 comes from the last terms.

$2(x^2 - 6x + 5) = 2(x \quad 5)(x \quad 1)$

Now we use the $-6x$, which comes from the outer and inner terms.

$2(x^2 - 6x + 5) = 2(x - 5)(x - 1)$

So $2x^2 - 12x + 10 = 2(x - 5)(x - 1)$.

EXAMPLE 13 Simplify $\dfrac{x^2 + 3x + 2}{3x + 6}$.

First, completely factor top and bottom as follows.

$$\dfrac{(x + 2)(x + 1)}{3(x + 2)}$$

But $\dfrac{x + 2}{x + 2}$ is 1, so we can cancel.

$$\dfrac{\overset{1}{(\cancel{x + 2})}(x + 1)}{3\underset{1}{(\cancel{x + 2})}}$$

is the same as $\dfrac{x + 1}{3}$

So $\dfrac{x^2 + 3x + 2}{3x + 6} = \dfrac{x + 1}{3}$.

Exercises

Multiply.

1. $(x + 1)(x + 2)$
2. $(x + 3)(x + 4)$
3. $(x + 2)(x + 3)$
4. $(x + 1)(x + 4)$
5. $(x + 5)(x + 2)$
6. $(x + 2)(x - 2)$
7. $(y + 5)(y - 3)$
8. $(a - 3)(a - 5)$
9. $(x - 1)(x + 1)$
10. $(x - 2)(x - 3)$

11. $\left(x - \dfrac{1}{3} \right)\left(x - \dfrac{2}{3} \right)$

12. $\left(y - \dfrac{1}{2} \right)\left(y - \dfrac{3}{2} \right)$

13. $(2x + 3)(x + 1)$
14. $(3x - 2)(x + 1)$
15. $(x - 4)(2x - 5)$
16. $(3x + 2)(4x + 3)$
17. $(2x + 1)(3x - 2)$
18. $(3x - 5)(5x - 3)$
19. $(2x + 3)(2x - 3)$
20. $(x + 2)^2$
21. $(x - 0.5)^2$
22. $(2x + 0.5)^2$
23. $(x + 2y)(x + y)$
24. $(a - b)(a - 3b)$
25. $(a - b)(a + b)$
26. $(3a + 5b)(3a - 5b)$

Factor completely.

27. $x^2 + 3x + 2$
28. $x^2 - 3x + 2$
29. $x^2 - 8x + 7$
30. $x^2 - 6x + 5$
31. $x^2 + 6x + 5$
32. $x^2 + 12x + 11$
33. $3x^2 + 12x + 9$
34. $2x^2 + 16x + 14$
35. $2x^2 + 4x + 2$
36. $x^3 + 2x^2 + x$

Simplify.

37. $\dfrac{x^2 + 4x + 3}{3x + 9}$

38. $\dfrac{x - 1}{x^2 - x}$

39. $\dfrac{x^2 + 4x + 3}{x^2 + 6x + 5}$

40. $\dfrac{x^2 - 8x + 7}{x^2 - 12x + 11}$

Answers to Exercises

① $(x+1)(x+2)$
$x^2 + 2x + 1x + 2$
$x^2 + 3x + 2$

② $(x+3)(x+4)$
$x^2 + 4x + 3x + 12$
$x^2 + 7x + 12$

③ $(x+2)(x+3)$
$x^2 + 3x + 2x + 6$
$x^2 + 5x + 6$

④ $(x+1)(x+4)$
$x^2 + 4x + 1x + 4$
$x^2 + 5x + 4$

⑤ $(x+5)(x+2)$
$x^2 + 2x + 5x + 10$
$x^2 + 7x + 10$

⑥ $(x+2)(x-2)$
$x^2 - 2x + 2x - 4$
$x^2 - 4$

⑦ $(y+5)(y-3)$
$y^2 - 3y + 5y - 15$
$y^2 + 2y - 15$

⑧ $(a-3)(a-5)$
$a^2 - 5a - 3a + 15$
$a^2 - 8a + 15$

⑨ $(x-1)(x+1)$
$x^2 + 1x - 1x - 1$
$x^2 - 1$

⑩ $(x-2)(x-3)$
$x^2 - 3x - 2x + 6$
$x^2 - 5x + 6$

⑪ $\left(x - \frac{1}{3}\right)\left(x - \frac{2}{3}\right)$
$x^2 - \frac{2}{3}x - \frac{1}{3}x + \frac{2}{9}$
$x^2 - x + \frac{2}{9}$

⑫ $\left(y - \frac{1}{2}\right)\left(y - \frac{3}{2}\right)$
$y^2 - \frac{3}{2}y - \frac{1}{2}y + \frac{3}{4}$
$y^2 - 2y + \frac{3}{4}$

⑬ $(2x+3)(x+1)$
$2x^2 + 2x + 3x + 3$
$2x^2 + 5x + 3$

⑭ $(3x-2)(x+1)$
$3x^2 + 3x - 2x - 2$
$3x^2 + x - 2$

15. $(x-4)(2x-5)$
$2x^2 - 5x - 8x + 20$
$2x^2 - 13x + 20$

16. $(3x+2)(4x+3)$
$12x^2 + 9x + 8x + 6$
$12x^2 + 17x + 6$

17. $(2x+1)(3x-2)$
$6x^2 - 4x + 3x - 2$
$6x^2 - x - 2$

18. $(3x-5)(5x-3)$
$15x^2 - 9x - 25x + 15$
$15x^2 - 34x + 15$

19. $(2x+3)(2x-3)$
$4x^2 - 6x + 6x - 9$
$4x^2 - 9$

20. $(x+2)(x+2)$
$x^2 + 2x + 2x + 4$
$x^2 + 4x + 4$

21. $(x-0.5)(x-0.5)$
$x^2 - 0.5x - 0.5x + 0.25$
$x^2 - x + 0.25$

22. $(2x+0.5)(2x+0.5)$
$4x^2 + x + x + 0.25$
$4x^2 + 2x + 0.25$

23. $(x+2y)(x+y)$
$x^2 + xy + 2xy + 2y^2$
$x^2 + 3xy + 2y^2$

24. $(a-b)(a-3b)$
$a^2 - 3ab - ab + 3b^2$
$a^2 - 4ab + 3b^2$

25. $(a-b)(a+b)$
$a^2 + ab - ab - b^2$
$a^2 - b^2$

26. $(3a+5b)(3a-5b)$
$9a^2 - 15ab + 15ab - 25b^2$
$9a^2 - 25b^2$

27. $x^2 + 3x + 2$
$(x+2)(x+1)$

28. $x^2 - 3x + 2$
$(x-2)(x-1)$

29. $x^2 - 8x + 7$
$(x-1)(x-7)$

30. $x^2 - 6x + 5$
$(x-1)(x-5)$

31. $x^2 + 6x + 5$
$(x+1)(x+5)$

(32.) $x^2 + 12x + 11$
$(x+1)(x+11)$

(33.) $3x^2 + 12x + 9$
$3(x+1)(x+3)$

(34.) $2x^2 + 16x + 14$
$2(x+1)(x+7)$

(35.) $2x^2 + 4x + 2$
$2(x+1)(x+1)$

(36.) $x^3 + 2x^2 + x$
$x(x^2 + 2x + 1)$
$x(x+1)(x+1)$

(37.) $\dfrac{x^2 + 4x + 3}{3x + 9} = \dfrac{(x+1)(x+3)}{3(x+3)} = \dfrac{x+1}{3}$

(38.) $\dfrac{x-1}{x^2 - x} = \dfrac{(x-1)}{x(x-1)} = \dfrac{1}{x}$

(39.) $\dfrac{x^2 + 4x + 3}{x^2 + 6x + 5} = \dfrac{(x+1)(x+3)}{(x+1)(x+5)} = \dfrac{x+3}{x+5}$

(40.) $\dfrac{x^2 - 8x + 7}{x^2 - 12x + 11} = \dfrac{(x-1)(x-7)}{(x-1)(x-11)} = \dfrac{x-7}{x-11}$

Additional Exercises

Multiply.
1. $(x + 3)(x + 8)$
2. $(x + 5)(x + 6)$
3. $(x + 3)(x + 5)$
4. $(x + 1)(x + 7)$
5. $(x + 5)(x - 5)$
6. $(x + 7)(x - 6)$
7. $(x - 2)(x + 1)$
8. $(x - 3)(x + 4)$
9. $(x + 4)(x - 4)$
10. $(x - 5)(x - 3)$

11. $\left(y - \dfrac{1}{4} \right)\left(y - \dfrac{3}{4} \right)$

12. $\left(a - \dfrac{2}{5} \right)\left(a - \dfrac{3}{5} \right)$

13. $(2x + 1)(x + 1)$
14. $(x + 1)(3x + 2)$
15. $(5x + 3)(x + 4)$
16. $(2x - 3)(3x + 2)$
17. $(3x - 2)(4x + 1)$
18. $(3x - 2)(5x - 2)$
19. $(2x - 3)(5x - 4)$
20. $(x - 1)^2$
21. $(x + 0.3)^2$
22. $(a - 0.2)^2$
23. $(x + y)(2x + y)$
24. $(a - b)(a + 3b)$
25. $(2c - d)(c + d)$
26. $(2a - 3b)(3a - 2b)$

Factor completely.
27. $x^2 + 8x + 7$
28. $x^2 - 2x + 1$
29. $x^2 - 12x + 11$
30. $x^2 - 14x + 13$
31. $x^2 + 14x + 13$
32. $x^2 + 18x + 17$
33. $5x^2 + 15x + 10$
34. $3x^2 + 9x + 6$
35. $x^3 - 6x^2 + 5x$
36. $2x^3 - 4x^2 + 2x$

Simplify.

37. $\dfrac{x - 2}{x^2 - x - 2}$

38. $\dfrac{x^2 + 4x - 5}{x^2 - 2x + 1}$

39. $\dfrac{x^2 - 2x - 3}{2x - 6}$

40. $\dfrac{2x^2 - 16x + 14}{3x^2 - 6x + 3}$

Lesson 3

More Factoring

In this lesson we are going to factor different kinds of trinomials by the same method we used in the last lesson.

EXAMPLE 1 Factor $x^2 + 5x + 6$.

There is no common monomial factor, so we go on to the next step, which is to look for binomial factors. As in the last lesson the x^2 comes from multiplying the first terms.

$$x^2 + 5x + 6 = (x \quad ?)(x \quad ?)$$

The +6 comes from multiplying the last terms, so the question marks must stand for two numbers which multiply to give +6. The possibilities are

$$6 = (6)(1) \text{ and } 6 = (3)(2)$$
$$6 = (-6)(-1) \text{ and } 6 = (-3)(-2)$$

The way we decide which possibilities work is by trying. So let's try.

$$x^2 + 5x + 6 \overset{?}{=} (x + 6)(x + 1)$$

$$x^2 + 5x + 6 \overset{?}{=} (x - 6)(x - 1)$$

$$x^2 + 5x + 6 \overset{?}{=} (x - 3)(x - 2)$$

$$x^2 + 5x + 6 \overset{?}{=} (x + 3)(x + 2)$$

The middle term, $+5x$, comes from the outer and inner terms. Let's try the possibilities.

$(x + 6)(x + 1)$ $\begin{array}{l} +6x \\ +x \end{array}$ } add up to $+7x$ not $+5x$

$(x - 6)(x - 1)$ $\begin{array}{l} -6x \\ -x \end{array}$ } add up to $-7x$ not $+5x$

$(x - 3)(x - 2)$ $\begin{array}{l} -3x \\ -2x \end{array}$ } add up to $-5x$ not $5x$

$(x + 3)(x + 2)$ $\begin{array}{l} +3x \\ +2x \end{array}$ } add up to $+5x$ Success!

So $x^2 + 5x + 6 = (x + 3)(x + 2)$. Now let's use the FOIL method of multiplying to check.

$$(x + 3)(x + 2) \overset{?}{=} x^2 + 2x + 3x + 6$$
$$= x^2 + 5x + 6 \qquad \underline{\text{It checks!}}$$

EXAMPLE 2   Factor $x^2 - 9x + 8$.

$$x^2 - 9x + 8 = (x \quad ?)(x \quad ?)$$

The $+ 8$ comes from the last terms, so here are the possibilities.

$$8 = (8)(1) \qquad\qquad 8 = (4)(2)$$
$$8 = (-8)(-1) \qquad 8 = (-4)(-2)$$

$$x^2 - 9x + 8 \overset{?}{=} (x + 8)(x + 1)$$

$$x^2 - 9x + 8 \overset{?}{=} (x - 8)(x - 1)$$

$$x^2 - 9x + 8 \overset{?}{=} (x + 4)(x + 2)$$

$$x^2 - 9x + 8 \overset{?}{=} (x - 4)(x - 2)$$

The middle term, $-9x$, comes from the outer and inner terms. Let's try the possibilities.

$$+8x$$
$(x + 8)(x + 1)$ } add up to $+9x$
$+x$ } not $-9x$

$$-8x$$
$(x - 8)(x - 1)$ } add up to $-9x$
$-x$ } Success!

So $x^2 - 9x + 8 = (x - 8)(x - 1)$.

Note: Once you find a combination that works, you've got it. There are no others that will work.

EXAMPLE 3 Factor $x^2 + 3x - 10$.

$$x^2 + 3x - 10 = (x \quad ?)(x \quad ?)$$

The -10 comes from the last terms so the possibilities are

$$-10 = (5)(-2) \quad -10 = (1)(-10)$$
$$-10 = (-5)(2) \quad -10 = (-1)(10)$$

Let's try them.

$$+5x$$
$(x + 5)(x - 2)$ } which add up to $3x$ ___Success!___
$-2x$

We got it the first time! We don't have to look any further. So $x^2 + 3x - 10 = (x + 5)(x - 2)$. Check it by multiplying.

Note: We could have factored $x^2 + 3x - 10$ as $(x - 2)(x + 5)$. You can write factors in any order.

EXAMPLE 4 Factor $3x^2 - 9x - 30$.

This time, there is a common monomial factor, 3.

$$3x^2 - 9x - 30 = 3(x^2 - 3x - 10)$$

Now let's factor what's inside the parentheses.

$$3(x^2 - 3x - 10) = 3(x \quad ?)(x \quad ?)$$

The -10 comes from the last terms, so the possibilities are

$$-10 = (5)(-2) \qquad -10 = (1)(-10)$$
$$-10 = (-5)(2) \qquad -10 = (-1)(10)$$

Let's try them.

$3(x + 5)(x - 2)$ } add up to $3x$
 not $-3x$ — $5x$, $-2x$

$3(x - 5)(x + 2)$ } add up to $-3x$
 Success! — $-5x$, $2x$

So $3x^2 - 9x - 30 = 3(x - 5)(x + 2)$.

EXAMPLE 5   Factor $x^2 + 5x + 3$.

There is no common monomial factor.

$$x^2 + 5x + 3 = (x \quad ?)(x \quad ?)$$

The $+3$ comes from multiplying the last terms, so the only possibilities are $3 = (3)(1)$ and $3 = (-3)(-1)$.

$$x^2 + 5x + 3 \overset{?}{=} (x + 3)(x + 1)$$
$$x^2 + 5x + 3 \overset{?}{=} (x - 3)(x - 1)$$

The middle term, $+5x$, comes from the outer and inner terms.

$(x + 3)(x + 1)$ } add up to $+4x$
 not $+5x$ — $+3x$, $+x$

$(x - 3)(x - 1)$ } add up to $-4x$
 not $+5x$ — $-3x$, $-x$

Neither possibility works. So we cannot factor $x^2 + 5x + 3$. That is, $x^2 + 5x + 3$ is prime.

EXAMPLE 6   Factor $x^2 - 10x + 25$.

There is no common monomial factor.

$$x^2 - 10x + 25 = (x \quad ?)(x \quad ?)$$

The $+25$ comes from the last terms, so here are the possibilities.

$$25 = (25)(1) \qquad\qquad 25 = (5)(5)$$
$$25 = (-25)(-1) \qquad 25 = (-5)(-5)$$

$$x^2 - 10x + 25 \overset{?}{=} (x + 25)(x + 1)$$
$$x^2 - 10x + 25 \overset{?}{=} (x - 25)(x - 1)$$
$$x^2 - 10x + 25 \overset{?}{=} (x + 5)(x + 5)$$
$$x^2 - 10x + 25 \overset{?}{=} (x - 5)(x - 5)$$

The middle term, $-10x$, comes from outer and inner terms.

$(x + 25)(x + 1)$ add up to $+26x$ not $-10x$ (outer $+25x$, inner $+1x$)

$(x - 25)(x - 1)$ add up to $-26x$ not $-10x$ (outer $-25x$, inner $-1x$)

$(x + 5)(x + 5)$ add up to $+10x$ not $-10x$ (outer $+5x$, inner $+5x$)

$(x - 5)(x - 5)$ add up to $-10x$ Success! (outer $-5x$, inner $-5x$)

So $x^2 - 10x + 25 = (x - 5)(x - 5)$, or $(x - 5)^2$. Check it by multiplying.

EXAMPLE 7 Factor $x^2 - 9$.

There is no common monomial factor. Think of $x^2 - 9$ as a trinomial.
Think: $x^2 - 9 = x^2 + 0x - 9$.

$$x^2 + 0x - 9 = (x \quad ?)(x \quad ?)$$

The -9 comes from multiplying the last terms, so the possibilities are

$$-9 = (3)(-3)$$
$$-9 = (1)(-9)$$
$$-9 = (-1)(9)$$

$$\left.\begin{array}{c} 3x \\ (x + 3)(x - 3) \\ -3x \end{array}\right\} \quad \text{add up to 0} \qquad \underline{\text{Success!}}$$

So $x^2 - 9 = (x + 3)(x - 3)$.

Note: The inners and outers must be opposites to add up to zero (the middle term).

EXAMPLE 8 Factor $x^2 - 16$.

There is no common monomial factor. Think: $x^2 - 16 = x^2 + 0x - 16$.

$$x^2 + 0x - 16 = (x \quad ?)(x \quad ?)$$

The -16 comes from multiplying the last terms. Since the inners and outers must be opposites to add up to zero (the middle term), we will try

$$-16 = (-4)(4)$$
$$(x - 4)(x + 4) \overset{?}{=} x^2 - 16$$

$$\left.\begin{array}{c} -4x \\ (x - 4)(x + 4) \\ +4x \end{array}\right\} \quad \text{adds up to 0} \qquad \underline{\text{Success!}}$$

So $x^2 - 16 = (x - 4)(x + 4)$. Check this by multiplying.

The last two problems were a little different from the rest. Let's look at more of this type.

$$x^2 - 1 = (x + 1)(x - 1)$$
$$x^2 - 4 = (x + 2)(x - 2)$$
$$x^2 - 9 = (x + 3)(x - 3)$$
$$x^2 - 16 = (x + 4)(x - 4)$$
$$x^2 - 25 = (x + 5)(x - 5)$$
$$x^2 - 36 = (x + 6)(x - 6)$$
$$x^2 - 49 = (x + 7)(x - 7)$$
$$x^2 - 64 = (x + 8)(x - 8)$$

x^2 is a square and each of the numbers 1, 4, 9, 16, ..., 64 is also a square. So each of the expressions $x^2 - 1$, $x^2 - 4$, $x^2 - 16$, ..., $x^2 - 64$ is the **difference of two squares**.

Using the pattern above we can factor each of the following.

$$x^2 - 100 = (x + 10)(x - 10)$$
$$x^2 - y^2 = (x + y)(x - y)$$
$$\triangle^2 - \square^2 = (\triangle + \square)(\triangle - \square)$$

EXAMPLE 9   Factor $2x^2 - 18$.

This time 2 is a common monomial factor.

$$2x^2 - 18 = 2(x^2 - 9)$$

Note: $x^2 - 9$ is a difference of two squares.

So $2(x^2 - 9) = 2(x + 3)(x - 3)$.

EXAMPLE 10   Factor $x^2 - 4y^2$.

There is no common monomial factor.

Note: $x^2 - 4y^2$ is the difference of two squares.

$$x^2 - 4y^2 = (x)^2 - (2y)^2$$

So $x^2 - 4y^2 = (x + 2y)(x - 2y)$. Check by multiplying.

EXAMPLE 11 Factor $x^2 - 5$.

5 is not a square, so this expression is not a difference of two squares. Let's try to factor it anyway. Think: $x^2 - 5 = x^2 + 0x - 5$.

$$x^2 - 5 = (x \quad ?)(x \quad ?)$$
$$- 5 = (+5)(-1) \text{ or } -5 = (-5)(1)$$

None of the possibilities works. We cannot factor $x^2 - 5$. So $x^2 - 5$ is prime.

EXAMPLE 12 Factor $x^2 + 9$.

There is no common monomial factor and this is not a difference of two squares. Think: $x^2 + 9 = x^2 + 0x + 9$.

$$x^2 + 0x + 9 = (x \quad ?)(x \quad ?)$$

The +9 comes from multiplying the last terms. The possibilities are

$$+9 = (9)(1) \qquad +9 = (3)(3)$$
$$+9 = (-9)(-1) \qquad +9 = (-3)(-3)$$

The middle term, $0x$, comes from the outer and inner terms.

None of the possibilities works. We cannot factor $x^2 + 9$. So $x^2 + 9$ is prime.

EXAMPLE 13 Simplify $\dfrac{x^2 - 16}{x^2 + 6x + 8}$.

Factor top and bottom completely.

Top $\qquad\qquad x^2 - 16 = (x - 4)(x + 4)$
Bottom $\quad x^2 + 6x + 8 = (x + 2)(x + 4)$

$$\frac{x^2 - 16}{x^2 + 6x + 8} = \frac{(x - 4)(x + 4)}{(x + 2)(x + 4)}$$

We can cancel identical factors.

$$\frac{x^2 - 16}{x^2 + 6x + 8} = \frac{(x - 4)(\cancel{x + 4})^{1}}{(x + 2)(\cancel{x + 4})_{1}}$$

So $\qquad \dfrac{x^2 - 16}{x^2 + 6x + 8} = \dfrac{x - 4}{x + 2}$.

Exercises

Factor completely.

1. $x^2 + 6x + 5$
2. $x^2 + 8x + 7$
3. $x^2 - 12x + 11$
4. $x^2 - 2x - 3$
5. $3x^2 - 6x + 3$
6. $ax^2 + 7ax + 10a$
7. $x^2 + 5x + 4$
8. $x^2 + 4x + 4$
9. $x^2 - 3x - 10$
10. $x^2 - 9x + 14$
11. $x^2 - 8x - 9$
12. $x^2 - 64$
13. $x^2 + 7x - 8$
14. $x^2 - 6x + 9$
15. $y^2 - 2y + 1$
16. $2y^2 - 2$
17. $y^2 - 4x^2$
18. $x^2 + 16x + 64$
19. $x^2 + 64$
20. $x^2 - 49$

Simplify.

21. $\dfrac{3x - 12}{x^2 - 16}$

22. $\dfrac{x^2 + 4x + 3}{x^2 + 6x + 5}$

23. $\dfrac{x^2 - 5x + 6}{x - 2}$

24. $\dfrac{x^2 - 2x - 8}{x^2 + x - 2}$

25. $\dfrac{2x^2 - 14x + 12}{3x^2 - 3}$

Answers to Exercises

1. $x^2 + 6x + 5$
$(x+1)(x+5)$

2. $x^2 + 8x + 7$
$(x+1)(x+7)$

3. $x^2 - 12x + 11$
$(x-1)(x-11)$

4. $x^2 - 2x - 3$
$(x+1)(x-3)$

5. $3x^2 - 6x + 3$
$3(x^2 - 2x + 1)$
$3(x-1)(x-1)$

6. $ax^2 + 7ax + 10a$
$a(x^2 + 7x + 10)$
$a(x+2)(x+5)$

7. $x^2 + 5x + 4$
$(x+4)(x+1)$

8. $x^2 + 4x + 4$
$(x+2)(x+2)$

9. $x^2 - 3x - 10$
$(x+2)(x-5)$

10. $x^2 - 9x + 14$
$(x-2)(x-7)$

11. $x^2 - 8x - 9$
$(x+1)(x-9)$

12. $x^2 - 64$
$(x+8)(x-8)$

13. $x^2 + 7x - 8$
$(x+8)(x-1)$

14. $x^2 - 6x + 9$
$(x-3)(x-3)$

15. $y^2 - 2y + 1$
$(y-1)(y-1)$

16. $2y^2 - 2$
$2(y^2 - 1)$
$2(y+1)(y-1)$

17. $y^2 - 4x^2$
$(y-2x)(y+2x)$

18. $x^2 + 16x + 64$
$(x+8)(x+8)$

(19.) $x^2 + 64$,
prime!

(20.) $x^2 - 49$
$(x+7)(x-7)$

(21.) $\dfrac{3x-12}{x^2-16} = \dfrac{3(x-4)}{(x-4)(x+4)} = \dfrac{3}{x+4}$

(22.) $\dfrac{x^2+4x+3}{x^2+6x+5} = \dfrac{(x+1)(x+3)}{(x+1)(x+5)} = \dfrac{x+3}{x+5}$

(23.) $\dfrac{x^2-5x+6}{x-2} = \dfrac{(x-2)(x-3)}{(x-2)} = \dfrac{x-3}{1} = x-3$

(24.) $\dfrac{x^2-2x-8}{x^2+x-2} = \dfrac{(x+2)(x-4)}{(x+2)(x-1)} = \dfrac{x-4}{x-1}$

(25.) $\dfrac{2x^2-14x+12}{3x^2-3} = \dfrac{2(x^2-7x+6)}{3(x^2-1)} = \dfrac{2(x-1)(x-6)}{3(x-1)(x+1)} = \dfrac{2(x-6)}{3(x+1)}$

Additional Exercises

Factor completly.

1. $x^2 + 2x + 1$
2. $x^2 + 3x + 2$
3. $x^2 - 4x + 3$
4. $x^2 - 18x + 17$
5. $x^2 + x - 2$
6. $x^2 - x - 2$
7. $5x^2 - 5x - 10$
8. $x^2 - 2x - 3$
9. $3x^2 + 6x - 9$
10. $x^2 + 5x + 6$
11. $x^2 - 7x + 6$
12. $x^2 - x + 6$
13. $2x^2 + 2x - 12$
14. $x^2 - 1$
15. $x^2 - 100$
16. $x^2 - 25y^2$
17. $2x^2 - 98$
18. $x^2 + 11x + 24$
19. $x^2 - 2x - 24$
20. $x^2 + 2x - 48$

Simplify.

21. $\dfrac{5x - 15}{x^2 - 9}$

22. $\dfrac{x - 3}{x^2 - 5x + 6}$

23. $\dfrac{x^2 + 10x + 25}{x^2 - 25}$

24. $\dfrac{x^2 + x - 2}{3x^2 + 12x + 9}$

25. $\dfrac{2x^2 - 2x - 24}{2x^2 - 18}$

Still More Factoring

In this lesson we are going to factor trinomials that start with terms like $2x^2$ where the 2 is not a common factor.

Let's multiply $2x + 1$ by $x + 1$.

$$(2x + 1)(x + 1) = 2x^2 + 3x + 1$$

Now let's start with $2x^2 + 3x + 1$ and factor it.

$$2x^2 + 3x + 1 = (?\quad ?)(?\quad ?)$$

The $2x^2$ must have come from multiplying the first terms.

$$2x^2 + 3x + 1 = (2x\quad ?)(x\quad ?)$$

In the same way, the $+1$ comes from multiplying the last terms.

$$2x^2 + 3x + 1 = (2x\quad ?)(x\quad ?)$$

45

The remaining question marks must stand for two numbers which multiply to give $+1$. The possibilities are $1 = (+1)(+1)$ and $1 = (-1)(-1)$. Let's try them.

$$2x^2 + 3x + 1 \overset{?}{=} (2x + 1)(x + 1)$$
$$2x^2 + 3x + 1 \overset{?}{=} (2x - 1)(x - 1)$$

The middle term, $3x$, comes from the outer and inner terms.

$$(2x + 1)(x + 1) \left.\begin{array}{c} +x \\ \\ +2x \end{array}\right\} \text{ add up to } 3x \quad \underline{\text{Success!}}$$

So $2x^2 + 3x + 1 = (2x + 1)(x + 1)$.

EXAMPLE 1   Factor $5x^2 - 6x + 1$ completely.

There is no common monomial factor.

$$5x^2 - 6x + 1 = (5x \quad ?)(x \quad ?)$$

The $+1$ comes from multiplying the last terms, so the only possibilities are

$$1 = (1)(1) \text{ and } 1 = (-1)(-1)$$
$$5x^2 - 6x + 1 \overset{?}{=} (5x + 1)(x + 1)$$
$$5x^2 - 6x + 1 \overset{?}{=} (5x - 1)(x - 1)$$

$$(5x + 1)(x + 1) \left.\begin{array}{c} +x \\ \\ +5x \end{array}\right\} \begin{array}{l} \text{adds up to } +6x \\ \underline{\text{not } -6x} \end{array} \qquad (5x - 1)(x - 1) \left.\begin{array}{c} -x \\ \\ -5x \end{array}\right\} \begin{array}{l} \text{adds up to } -6x \\ \underline{\text{Success!}} \end{array}$$

So $5x^2 - 6x + 1 = (5x - 1)(x - 1)$.

EXAMPLE 2 Factor $2x^2 + 7x + 5$ completely.

There is no common monomial factor.

$$2x^2 + 7x + 5 = (2x \quad ?)(x \quad ?)$$

The $+5$ comes from multiplying the last terms, so the possibilities are

$$5 = (1)(5) \text{ and } 5 = (-1)(-5)$$

But in this kind of problem, we must also try the possibilities in reverse order, because they will give us different middle terms.

$$5 = (5)(1) \text{ and } 5 = (-5)(-1)$$
$$2x^2 + 7x + 5 \stackrel{?}{=} (2x + 1)(x + 5)$$
$$2x^2 + 7x + 5 \stackrel{?}{=} (2x + 5)(x + 1)$$
$$2x^2 + 7x + 5 \stackrel{?}{=} (2x - 1)(x - 5)$$
$$2x^2 + 7x + 5 \stackrel{?}{=} (2x - 5)(x - 1)$$

$(2x + 1)(x + 5)$ $\left.\begin{array}{c}\end{array}\right\}$ adds up to $11x$ not $7x$

$(2x + 5)(x + 1)$ $\left.\begin{array}{c}\end{array}\right\}$ adds up to $7x$ Success!

So $2x^2 + 7x + 5 = (2x + 5)(x + 1)$.

EXAMPLE 3 Factor $3x^2 - 7x + 4$ completely.

There is no common monomial factor.

$$3x^2 - 7x + 4 = (3x \quad ?)(x \quad ?)$$

The $+4$ comes from multiplying the last terms, so the possibilities are

$$4 = (2)(2) \qquad 4 = (1)(4)$$
$$4 = (-2)(-2) \qquad 4 = (-1)(-4)$$

But since $3x$ and x are different, we must consider two more possibilities.

$$4 = (4)(1)$$
$$4 = (-4)(-1)$$

Let's try them.

$$
\underbrace{(3x + 2)(x + 2)}_{\substack{\overset{\displaystyle -2x}{} \\ \displaystyle -6x}} \left\}\; \begin{array}{l} \text{adds up to } 8x \\ \underline{\text{not } -7x} \end{array} \right.
\qquad
\underbrace{(3x - 2)(x - 2)}_{\substack{\overset{\displaystyle -2x}{} \\ \displaystyle -6x}} \left\}\; \begin{array}{l} \text{adds up to } -8x \\ \underline{\text{not } -7x} \end{array} \right.
$$

$$
\underbrace{(3x + 1)(x + 4)}_{\substack{\overset{\displaystyle x}{} \\ \displaystyle 12x}} \left\}\; \begin{array}{l} \text{adds up to } 13x \\ \underline{\text{not } -7x} \end{array} \right.
\qquad
\underbrace{(3x - 1)(x - 4)}_{\substack{\overset{\displaystyle -x}{} \\ \displaystyle -12x}} \left\}\; \begin{array}{l} \text{adds up to } -13x \\ \underline{\text{not } -7x} \end{array} \right.
$$

$$
\underbrace{(3x + 4)(x + 1)}_{\substack{\overset{\displaystyle 4x}{} \\ \displaystyle 3x}} \left\}\; \begin{array}{l} \text{adds up to } 7x \\ \underline{\text{not } -7x} \end{array} \right.
\qquad
\underbrace{(3x - 4)(x - 1)}_{\substack{\overset{\displaystyle -4x}{} \\ \displaystyle -3x}} \left\}\; \begin{array}{l} \text{adds up to } -7x \\ \underline{\text{Success!}} \end{array} \right.
$$

So $3x^2 - 7x + 4 - (3x - 4)(x - 1)$.

We found the right possibility on the last try. With practice, you may be able to select the right possibility on an earlier try.

EXAMPLE 4   Factor $6x^2 + 16x + 8$ completely.

There is a common monomial factor, 2. Let's factor out the 2.

$$6x^2 + 16x + 8 = 2(3x^2 + 8x + 4)$$

Now let's try to factor the trinomial inside the parentheses.

$$2(3x^2 + 8x + 4) = 2(3x \quad ?)(x \quad ?)$$

The $+4$ comes from multiplying the last terms, so the possibilities are

$$4 = (2)(2) \qquad 4 = (1)(4)$$
$$4 = (-2)(-2) \qquad 4 = (-1)(-4)$$
$$ \qquad 4 = (4)(1)$$
$$ \qquad 4 = (-4)(-1)$$

Let's try them.

So $6x^2 + 16x + 8 = 2(3x + 2)(x + 2)$.

See: Experience helps to find the right possibilities sooner, or so we hope, anyway!

EXAMPLE 5   Factor $4y^2 + 7y - 2$ completely.

There is no common monomial factor.

$$4y^2 + 7y - 2 = (?\quad ?)(?\quad ?)$$

There are two possibilities for the first terms since

$$4y^2 = (2y)(2y) \text{ and } 4y^2 = (4y)(y)$$

Since there are two possibilities for the first terms, we have to break the problem into two separate parts.

$$4y^2 + 7y - 2 = (2y\quad ?)(2y\quad ?)$$
$$\text{or}\quad 4y^2 + 7y - 2 = (4y\quad ?)(y\quad ?)$$

The -2 comes from multiplying the last terms, so the possibilities are

$$-2 = (-2)(+1) \qquad -2 = (+2)(-1)$$
$$-2 = (+1)(-2) \qquad -2 = (-1)(+2)$$

Here we try $(2y\quad ?)(2y\quad ?)$ Here we try $(4y\quad ?)(y\quad ?)$

$(2y - 2)(2y + 1)\Big\}$ adds up to $-2y$ not $7y$

$(4y - 2)(y + 1)\Big\}$ adds up to $2y$ not $7y$

$(2y + 1)(2y - 2)$: Since the <u>first terms are the same, reversing</u> the order of the last terms does not change the middle term.

$(4y + 1)(y - 2)\Big\}$ adds up to $-7y$ not $7y$

$(2y + 2)(2y - 1)\Big\}$ adds up to $2y$ not $7y$

$(4y + 2)(y - 1)\Big\}$ adds up to $-2y$ not $7y$

$(2y - 1)(2y + 2)$: Since the first terms are the same, reversing the order of the last terms does not change the middle term.

$(4y - 1)(y + 2)\Big\}$ adds up to $7y$ Success!

So $4y^2 + 7y - 2 = (4y - 1)(y + 2)$.

EXAMPLE 6   Factor $8s^2 - 10s + 3$ completely.

There is no common monomial factor. There are two possibilities for the first terms.

$$8s^2 = (4s)(2s) \text{ and } 8s^2 = (8s)(s)$$

$$8s^2 - 10s + 3 = (4s\quad ?)(2s\quad ?)$$
$$\text{or}\quad 8s^2 - 10s + 3 = (8s\quad ?)(s\quad ?)$$

The $+3$ comes from multiplying the last terms, so the possibilities are

$$3 = (3)(1)\qquad 3 = (-3)(-1)$$
$$3 = (1)(3)\qquad 3 = (-1)(-3)$$

So the possible factors, then, are

$(4s + 3)(2s + 1)$ $(8s + 3)(s + 1)$
$(4s + 1)(2s + 3)$ $(8s + 1)(s + 3)$

$\boxed{(4s - 3)(2s - 1)}$ $(8s - 3)(s - 1)$

$(4s - 1)(2s - 3)$ $(8s - 1)(s - 3)$

These problems can get ridiculous! The one with the box around it works. So $8s^2 - 10s + 3 = (4s - 3)(2s - 1)$.

Remember, with practice you will be able to eliminate some of the possibilities without trying them by looking at the signs of the original trinomial.

EXAMPLE 7   Simplify $\dfrac{4x^2 - 14x - 30}{2x^2 + 13x + 15}$.

Factor top and bottom completely.

Top $4x^2 - 14x - 30 = 2(2x^2 - 7x - 15)$

We can factor the top further.

$$4x^2 - 14x - 30 = 2(2x + 3)(x - 5)$$

Bottom $2x^2 + 13x + 15 = (2x + 3)(x + 5)$

So $\dfrac{4x^2 - 14x - 30}{2x^2 + 13x + 15} = \dfrac{2(2x + 3)(x - 5)}{(2x + 3)(x + 5)}$.

We can cancel identical factors.

$$\dfrac{4x^2 - 14x - 30}{2x^2 + 13x + 15} = \dfrac{2(\cancel{2x + 3})^1(x - 5)}{(\cancel{2x + 3})_1(x + 5)}$$

So $\dfrac{4x^2 - 14x - 30}{2x^2 + 13x + 15} = \dfrac{2(x - 5)}{x + 5}$.

Exercises

Factor completely.

1. $2x^2 - 3x + 1$
2. $11x^2 + 12x + 1$
3. $11x^2 - 12x + 1$
4. $2x^2 - x - 1$
5. $2x^2 + 4x + 2$
6. $2x^2 + 5x + 2$
7. $2x^2 - x - 3$
8. $2x^2 + 5x + 6$
9. $3y^2 - 4y - 4$
10. $3x^3 - 5x^2 - 12x$
11. $5x^2y - 17xy + 6y$
12. $6x^2 - 11x + 5$
13. $49x^2 - 36$
14. $7x^2 + 9x - 10$
15. $4z^2 - 16z + 7$

Simplify.

16. $\dfrac{2x^2 - x - 1}{x^2 - 1}$

17. $\dfrac{6x^2 - 7x - 3}{8x - 12}$

18. $\dfrac{9x^2 - 49}{3x^2 + 10x + 7}$

19. $\dfrac{(2x + 3)^2}{2x^2 - 7x - 15}$

20. $\dfrac{6x + 8}{3x^2 + 10x + 8}$

Answers to Exercises

1. $2x^2 - 3x + 1$

$(2x - 1)(x - 1)$

2. $11x^2 + 12x + 1$

$(11x + 1)(x + 1)$

3. $11x^2 - 12x + 1$

$(11x - 1)(x - 1)$

4. $2x^2 - x - 1$

$(2x + 1)(x - 1)$

5. $2x^2 + 4x + 2$
$2(x^2 + 2x + 1)$

$2(x + 1)(x + 1)$

6. $2x^2 + 5x + 2$

$(2x + 1)(x + 2)$

7. $2x^2 - x - 3$

$(2x - 3)(x + 1)$

8. $2x^2 + 5x + 6$

prime!

9. $3y^2 - 4y - 4$

$(3y + 2)(y - 2)$

10. $3x^3 - 5x^2 - 12x$

$x(3x^2 - 5x - 12)$

$x(3x + 4)(x - 3)$

11. $5x^2y - 17xy + 6y$

$y(5x^2 - 17x + 6)$

$y(5x - 2)(x - 3)$

12. $6x^2 - 11x + 5$

$(6x - 5)(x - 1)$

13. $49x^2 - 36$

$(7x - 6)(7x + 6)$

14. $7x^2 + 9x - 10$

$(7x - 5)(x + 2)$

(15.) $4z^2 - 16z + 7$

$(2z-7)(2z-1)$

(16.) $\dfrac{2x^2 - x - 1}{x^2 - 1} = \dfrac{(2x+1)(x-1)}{(x+1)(x-1)} = \dfrac{2x+1}{x+1}$

(17.) $\dfrac{6x^2 - 7x - 3}{8x - 12} = \dfrac{(2x-3)(3x+1)}{4(2x-3)} = \dfrac{3x+1}{4}$

(18.) $\dfrac{9x^2 - 49}{3x^2 + 10x + 7} = \dfrac{(3x+7)(3x-7)}{(3x+7)(x+1)} = \dfrac{3x-7}{x+1}$

(19.) $\dfrac{(2x+3)^2}{2x^2 - 7x - 15} = \dfrac{(2x+3)(2x+3)}{(2x+3)(x-5)} = \dfrac{2x+3}{x-5}$

(20.) $\dfrac{6x+8}{3x^2 + 10x + 8} = \dfrac{2(3x+4)}{(3x+4)(x+2)} \qquad \dfrac{2}{x+2}$

Additional Exercises

Factor completely.

1. $5x^2 + 6x + 1$
2. $7x^2 - 6x - 1$
3. $3y^2 - 4y + 1$
4. $11y^2 - 10y - 1$
5. $50y^2 + 20y + 2$
6. $3x^3 - 14x^2 - 5x$
7. $2a^2 - 9a + 10$
8. $2a^2 - 21a + 10$
9. $2a^2 + a - 10$
10. $14a^2 - 14$
11. $3b^2 + 7b - 6$
12. $3b^2 + 17b - 6$
13. $3b^2 - 5b - 8$
14. $10x^2 + 17x + 3$
15. $10x^2 - 21x + 2$

Simplify.

16. $\dfrac{3x^2 - 14x - 5}{3x^2 - 2x - 1}$

17. $\dfrac{6x^2 - 13x - 5}{6x^2 - 15x}$

18. $\dfrac{25x^2 - 100}{75x^2 + 100x - 100}$

19. $\dfrac{7x^2 - x}{7x^3 - 22x^2 + 3x}$

20. $\dfrac{4x^2 - 1}{4x^2 - 4x + 1}$

You Guessed It! Factoring

In this lesson we will factor trinomials like $x^2 + 3xy + 2y^2$ which contain two different letters. We can factor such trinomials using the same methods as in previous lessons.

EXAMPLE 1 Factor completely: $x^2 + 3xy + 2y^2$

There is no common monomial factor.

$$x^2 + 3xy + 2y^2 = (x \quad ?y)(x \quad ?y)$$

The $2y^2$ comes from multiplying the last terms, so the possibilities are

$$2y^2 = (2y)(y)$$
or $\quad 2y^2 = (-2y)(-y)$

Let's try them.

$$- 2xy$$

$$(x + 2y)(x + y) \Big\} \quad \text{add up to } 3xy$$

$$\underline{\text{Success!}}$$

$$xy$$

So $x^2 + 3xy + 2y^2 = (x + 2y)(x + y)$.

EXAMPLE 2   Factor $6x^2 - 13xy + 5y^2$ completely.

There is no common monomial factor.

$$6x^2 - 13xy + 5y^2 = (?x \quad ?y)(?x \quad ?y)$$

The $6x^2$ comes from multiplying the first terms, so the possibilities are

$$6x^2 = (6x)(x)$$
or $\quad 6x^2 = (2x)(3x)$
$$6x^2 - 13xy + 5y^2 = (6x \quad ?y)(x \quad ?y)$$
or $\quad 6x^2 - 13xy + 5y^2 = (2x \quad ?y)(3x \quad ?y)$

The $+5y^2$ comes from multiplying the last terms, so the possibilities are

$$+5y^2 = (+5y)(+y) \qquad +5y^2 = (-5y)(-y)$$
$$+5y^2 = (+y)(+5y) \qquad +5y^2 = (-y)(-5y)$$

Let's try them, and see if we can find one that gives the correct middle term.

Here we try $(6x \quad ?y)(x \quad ?y)$ Here we try $(2x \quad ?y)(3x \quad ?y)$

$$- 5xy \Big\} \quad \text{adds up to}$$
$$(6x + 5y)(x + y) \quad 11xy \text{ not}$$
$$6xy \Big\} \quad -13xy$$

$$-15xy \Big\} \quad \text{adds up to}$$
$$(2x + 5y)(3x + y) \quad 17xy \text{ not}$$
$$2xy \Big\} \quad -13xy$$

$(6x + y)(x + 5y)$ — xy and $30xy$ — adds up to $31xy$ <u>not</u> $-13xy$

$(2x + y)(3x + 5y)$ — $3xy$ and $10xy$ — adds up to $13xy$ <u>not</u> $-13xy$

$(6x - 5y)(x - y)$ — $-5xy$ and $-6xy$ — adds up to $-11xy$ <u>not</u> $-13xy$

$(2x - 5y)(3x - y)$ — $-15xy$ and $-2xy$ — adds up to $-17xy$ <u>not</u> $-13xy$

$(6x - y)(x - 5y)$ — $-xy$ and $-30xy$ — adds up to $-31xy$ <u>not</u> $-13xy$

$(2x - y)(3x - 5y)$ — $-3xy$ and $-10xy$ — adds up to $-13xy$ <u>Success!</u>

So $6x^2 - 13xy + 5y^2 = (2x - y)(3x - 5y)$.

EXAMPLE 3 Factor completely: $4x^2 - 0.09y^2$

Note: There is no middle term.

> $4x^2$ is the square of $2x$
> $0.09y^2$ is the square of $0.3y$
> $4x^2 - 0.09y^2$ is the difference of two squares.

So $4x^2 - 0.09y^2 = (2x + 0.3y)(2x - 0.3y)$. Now let's use the FOIL method of multiplying to check.

$$(2x + 0.3y)(2x - 0.3y) \stackrel{?}{=} 4x^2 - 0.6xy + 0.6xy - 0.09y^2$$
$$= 4x^2 - 0.09y^2 \quad \underline{\text{It Checks!}}$$

EXAMPLE 4   Simplify $\dfrac{r^2 - 3rs + 2s^2}{2r^2 - rs - s^2}$.

First, factor top and bottom completely.

Top $r^2 - 3rs + 2s^2 = (r - s)(r - 2s)$

Bottom $2r^2 - rs - s^2 = (r - s)(2r + s)$

So $\dfrac{r^2 - 3rs + 2s^2}{2r^2 - rs - s^2} = \dfrac{(r - s)(r - 2s)}{(r - s)(2r + s)}$

We can cancel identical top and bottom factors.

$$\dfrac{r^2 - 3rs + 2s^2}{2r^2 - rs - s^2} = \dfrac{\overset{1}{\cancel{(r - s)}}(r - 2s)}{\underset{1}{\cancel{(r - s)}}(2r + s)}$$

So $\dfrac{r^2 - 3rs + 2s^2}{2r^2 - rs - s^2} = \dfrac{r - 2s}{2r + s}$.

Fractions with Opposites

Whenever the top and bottom of a fraction are exactly the same, we cancel and get $\dfrac{1}{1}$ or 1. Let's see what happens when the top and bottom are *opposites* of each other. Numbers or expressions are opposites when their sum is zero. This is illustrated below.

-6 and $+6$ are opposites and $\dfrac{-6}{+6} = -1$

x and $-x$ are opposites and $\dfrac{x}{-x} = -1$

$-3y$ and $3y$ are opposites and $\dfrac{-3y}{3y} = -1$

When the top and bottom of a fraction are opposites, the fraction equals -1.

EXAMPLE 5   Simplify $\dfrac{5x^2y^3}{-5x^2y^3}$.

The top and bottom of this fraction are opposites, so $\dfrac{5x^2y^3}{-5x^2y^3} = -1$.

EXAMPLE 6 What is the opposite of $3r^3 - 2r^2 + 3r + 1$?

This expression and its opposite must add to zero. Since $(3r^3 - 2r^2 + 3r + 1) + (-3r^3 + 2r^2 - 3r - 1) = 0$, the opposite of $3r^3 - 2r^2 + 3r + 1$ is $-3r^3 + 2r^2 - 3r - 1$.

Note: In example 6, <u>each term</u> of the polynomial $3r^3 - 2r^2 + 3r + 1$ is the opposite of a term of the polynomial $-3r^3 + 2r^2 - 3r - 1$. This gives us a test for opposite polynomials: If each term of one polynomial is the opposite of a term of another polynomial, the polynomials are opposites of each other.

EXAMPLE 7 Simplify $\dfrac{2x - 5}{-2x + 5}$.

Each term of $2x - 5$ is the opposite of a term of $-2x + 5$, so the top and bottom of this fraction are opposites.

So $\dfrac{2x - 5}{-2x + 5} = -1$.

EXAMPLE 8 Simplify $\dfrac{3a^2 - 2ab + b}{-3a^2 + 2ab - b}$.

Each term of the top is the opposite of a term of the bottom, so the top and bottom of this fraction are opposites. Thus, $\dfrac{3a^2 - 2ab + b}{-3a^2 + 2ab - b} = -1$.

EXAMPLE 9 Simplify $\dfrac{x^2 - xy}{y^2 - xy}$.

1. First, factor top and bottom as follows.

$$\frac{x^2 - xy}{y^2 - xy} = \frac{x(x - y)}{y(y - x)}$$

2. Now, notice that $(x - y)$ and $(y - x)$ are opposites.

So $\dfrac{x - y}{y - x} = -1$

and $\dfrac{x(x - y)}{y(y - x)}$ is $\dfrac{x}{y}(-1)$ or $-\dfrac{x}{y}$

So $\dfrac{x^2 - xy}{y^2 - xy} = -\dfrac{x}{y}$.

EXAMPLE 10   Simplify $\dfrac{x^2 - 4y^2}{2y^2 + xy - x^2}$.

First, factor top and bottom as follows.

$$\frac{x^2 - 4y^2}{2y^2 + xy - x^2} = \frac{(x - 2y)(x + 2y)}{(2y - x)(y + x)}$$

Notice that $x - 2y$ and $2y - x$ are opposites.

So $\dfrac{x - 2y}{2y - x} = -1$

and $\dfrac{(x - 2y)(x + 2y)}{(2y - x)(y + x)} = (-1)\dfrac{(x + 2y)}{y + x}$

or $\dfrac{-x - 2y}{y + x}$

So $\dfrac{x^2 - 4y^2}{2y^2 + xy - x^2} = \dfrac{-x - 2y}{y + x}$.

Exercises

Factor completely.

1. $x^2 + 2xy + y^2$
2. $x^2 - 2xy + y^2$
3. $x^2 - 3xy + 2y^2$
4. $x^2 - 0.25y^2$
5. $x^2 - 0.04y^2$
6. $x^2 + xy - 6y^2$
7. $x^2 - xy - 2y^2$
8. $2x^2 + 3xy + y^2$
9. $6x^2 + 7xy - 3y^2$
10. $8x^2 - 10xy - 3y^2$

Simplify.

11. $\dfrac{x + 3y}{x^2 - 9y^2}$

12. $\dfrac{2x + 4y}{x^2 + 4xy + 4y^2}$

13. $\dfrac{x^2 - 6xy + 9y^2}{(x - 3y)^2}$

14. $\dfrac{xy - 3y^2}{x^2 - xy - 6y^2}$

15. $\dfrac{x^2 - 8xy + 15y^2}{2x - 10y}$

16. $\dfrac{(x - 0.4y)^2}{x^2 - 0.16y^2}$

17. $\dfrac{3ab - 3a^2}{6b^2 - 6a^2}$

18. $\dfrac{x^2 - 16y^2}{x^2 - 8xy + 16y^2}$

19. $\dfrac{2r^2 - rs - 6s^2}{3r^2 - 12s^2}$

20. $\dfrac{2r^2 - 7rs + 6s^2}{2r^2 - 3rs - 2s^2}$

21. $\dfrac{+14a^2b}{-14a^2b}$

22. $\dfrac{21 - 17}{17 - 21}$

23. $\dfrac{x + 2y}{-x - 2y}$

24. $\dfrac{x - 2y}{-x + 2y}$

25. $\dfrac{x - y}{y - x}$

26. $\dfrac{x^2 - xy}{y - x}$

27. $\dfrac{x^2 - 4}{(2 - x)^2}$

28. $\dfrac{2b^2 - 3ab + a^2}{a - 2b}$

29. $\dfrac{y^2 - x^2}{x^2 - 2xy + y^2}$

30. $\dfrac{x^2 - 2xy - 3y^2}{9y - 3x}$

Answers to Exercises

① $x^2 + 2xy + y^2$

$(x+y)(x+y)$

⑥ $x^2 + xy - 6y^2$

$(x-2y)(x+3y)$

② $x^2 - 2xy + y^2$

$(x-y)(x-y)$

⑦ $x^2 - xy - 2y^2$

$(x+y)(x-2y)$

③ $x^2 - 3xy + 2y^2$

$(x-y)(x-2y)$

⑧ $2x^2 + 3xy + y^2$

$(2x+y)(x+y)$

④ $x^2 - 0.25y^2$

$(x+0.5y)(x-0.5y)$

⑨ $6x^2 + 7xy - 3y^2$

$(3x-y)(2x+3y)$

⑤ $x^2 - 0.04y^2$

$(x+0.2y)(x-0.2y)$

⑩ $8x^2 - 10xy - 3y^2$

$(4x+y)(2x-3y)$

⑪ $\dfrac{x+3y}{x^2-9y^2} = \dfrac{\cancel{(x+3y)}}{\cancel{(x+3y)}(x-3y)} = \dfrac{1}{x-3y}$

⑫ $\dfrac{2x+4y}{x^2+4xy+4y^2} = \dfrac{2\cancel{(x+2y)}}{\cancel{(x+2y)}(x+2y)} = \dfrac{2}{x+2y}$

⑬ $\dfrac{x^2-6xy+9y^2}{(x-3y)^2} = \dfrac{\cancel{(x-3y)}\cancel{(x-3y)}}{\cancel{(x-3y)}\cancel{(x-3y)}} = \dfrac{1}{1} = 1$

$$\text{(14.)} \quad \frac{xy - 3y^2}{x^2 - xy - 6y^2} = \frac{y\,\cancel{(x - 3y)}}{\cancel{(x - 3y)}\,(x + 2y)} = \frac{y}{x + 2y}$$

$$\text{(15.)} \quad \frac{x^2 - 8xy + 15y^2}{2x - 10y} = \frac{\cancel{(x - 5y)}(x - 3y)}{2\cancel{(x - 5y)}} = \frac{x - 3y}{2}$$

$$\text{(16.)} \quad \frac{(x - 0.4y)^2}{x^2 - 0.16y^2} = \frac{\cancel{(x - 0.4y)}(x - 0.4y)}{\cancel{(x - 0.4y)}(x + 0.4y)} = \frac{x - 0.4y}{x + 0.4y}$$

$$\text{(17.)} \quad \frac{3ab - 3a^2}{6b^2 - 6a^2} = \frac{3a\,\cancel{(b - a)}}{6\cancel{(b - a)}(b + a)} = \frac{\cancel{3}a}{\cancel{3}(2)(b + a)} = \frac{a}{2(b + a)}$$

$$\text{(18.)} \quad \frac{x^2 - 16y^2}{x^2 - 8xy + 16y^2} = \frac{\cancel{(x - 4y)}(x + 4y)}{\cancel{(x - 4y)}(x - 4y)} = \frac{x + 4y}{x - 4y}$$

$$\text{(19.)} \quad \frac{2r^2 - rs - 6s^2}{3r^2 - 12s^2} = \frac{(2r + 3s)\cancel{(r - 2s)}}{3\cancel{(r - 2s)}(r + 2s)} = \frac{2r + 3s}{3(r + 2s)}$$

$$\text{(20.)} \quad \frac{2r^2 - 7rs + 6s^2}{2r^2 - 3rs - 2s^2} = \frac{(2r - 3s)\cancel{(r - 2s)}}{(2r + s)\cancel{(r - 2s)}} = \frac{2r - 3s}{2r + s}$$

Answers 21-30 are continued on the next page.

21. $\dfrac{+14\,a^2\,b}{-14\,a^2\,b} = -1$

22. $\dfrac{21-17}{17-21} = -1$

notice that these are all -1, since the top and bottom are opposites.

23. $\dfrac{x+2y}{-x-2y} = -1$

24. $\dfrac{x-2y}{-x+2y} = -1$

25. $\dfrac{x-y}{y-x} = -1$

26. $\dfrac{x^2-xy}{y-x} = \dfrac{x(x-y)}{(y-x)} = x(-1) = -1x = -x$

27. $\dfrac{x^2-4}{(2-x)^2} = \dfrac{(x-2)(x+2)}{(2-x)(2-x)} = \dfrac{-1(x+2)}{2-x} = \dfrac{-x-2}{2-x}$

28. $\dfrac{2b^2-3ab+a^2}{a-2b} = \dfrac{(2b-a)(b-a)}{(a-2b)} = -1(b-a) = -b+a$

29. $\dfrac{y^2-x^2}{x^2-2xy+y^2} = \dfrac{(y-x)(y+x)}{(x-y)(x-y)} = \dfrac{-1(y+x)}{(x-y)} = \dfrac{-y-x}{x-y}$

30. $\dfrac{x^2-2xy-3y^2}{9y-3x} = \dfrac{(x-3y)(x+y)}{3(3y-x)} = \dfrac{-1(x+y)}{3} = \dfrac{-x-y}{3}$

Additional Exercises

Factor completely.
1. $x^2 + 3xy + 2y^2$
2. $x^2 - 4xy + 3y^2$
3. $x^2 - 9y^2$
4. $x^2 - 0.16$
5. $4x^2 - 0.09$
6. $x^2 + 4xy + 4y^2$
7. $x^2 + xy - 2y^2$
8. $2x^2 - 3xy + y^2$
9. $3x^2 - xy - 4y^2$
10. $6x^2 - 11xy + 5y^2$

Simplify.
11. $\dfrac{x^2 - y^2}{x - y}$

12. $\dfrac{x^2 + xy - 12y^2}{x + 4y}$

13. $\dfrac{x^2 - xy - 12y^2}{2x - 8y}$

14. $\dfrac{xy + 2y^2}{x^2 - 2xy - 8y^2}$

15. $\dfrac{x^2 - 7xy + 6y^2}{2x - 12y}$

16. $\dfrac{x^2 - 0.25y^2}{(x - 0.5y)^2}$

17. $\dfrac{8c^2 - 8d^2}{4cd - 4d^2}$

18. $\dfrac{x^2 - 36y^2}{x^2 - 12xy + 36y^2}$

19. $\dfrac{4r^2 - 9t^2}{2r^2 - 5rt + 3t^2}$

20. $\dfrac{3a^2 - 7ab + 4b^2}{3a^2 - 2ab - b^2}$

21. $\dfrac{15x^2y^3}{-15x^2y^3}$

22. $\dfrac{8 - 13}{13 - 8}$

23. $\dfrac{y - 7x}{-y + 7x}$

24. $\dfrac{x - y}{2y - 2x}$

25. $\dfrac{8 - 2x}{x^2 - 16}$

26. $\dfrac{xy - x^2}{y^2 - x^2}$

27. $\dfrac{x^2 - 25}{(5 - x)^2}$

28. $\dfrac{3x^2 - 4xy + y^2}{y - 3x}$

29. $\dfrac{a^2 - 4ab + 3b^2}{b^2 - a^2}$

30. $\dfrac{4y - 2x}{x^2 - xy - 2y^2}$

Lesson 6

Multiplication and Division of Algebraic Fractions

In this lesson we will multiply and divide algebraic fractions, and write the answer in simplest form.

Multiplication of Algebraic Fractions

Recall how you multiply fractions.

$$\left(\frac{3}{4}\right)\left(\frac{7}{5}\right) = \frac{(3)(7)}{(4)(5)} \quad \text{or} \quad \frac{21}{20}$$

If there is a factor which is both on the top and on the bottom, you can save yourself some work by canceling before you multiply. For example,

$$\left(\frac{3}{4}\right)\left(\frac{2}{3}\right) = \frac{3}{(2)(2)} \cdot \frac{2}{3}$$

Now we can cancel.

$$\frac{\cancel{3}^{\;1}}{(2)(\cancel{2})} \cdot \frac{\cancel{2}^{\;1}}{\cancel{3}}$$
$$\phantom{\frac{}{(2)(2)}}{}_{1}{}_{1}$$

is the same as $\dfrac{1}{2}$

So $\qquad \left(\dfrac{3}{4}\right)\left(\dfrac{2}{3}\right) = \dfrac{1}{2}$.

EXAMPLE 1 Multiply $\dfrac{a}{b}$ by $\dfrac{c}{d}$.

$$\frac{a}{b} \cdot \frac{c}{d} = \frac{ac}{bd}$$

EXAMPLE 2 Multiply $\dfrac{5b}{a^2}$ by $\dfrac{a}{b^4}$ and simplify.

$$\frac{5b}{a^2} \cdot \frac{a}{b^4}$$

Factor first.

$$\frac{5b}{aa} \cdot \frac{a}{bbbb}$$

Now cancel.

$$\frac{5\cancel{b}^{\;1}}{a\cancel{a}} \cdot \frac{\cancel{a}^{\;1}}{bbb\cancel{b}} = \frac{5}{ab^3}$$
$$\phantom{\frac{5b}{a}}{}_{1}{}_{1}$$

So $\qquad \dfrac{5b}{a^2} \cdot \dfrac{a}{b^4} = \dfrac{5}{ab^3}$,

EXAMPLE 3   Multiply $\dfrac{5a + 25}{2a}$ by $\dfrac{4a}{2a + 10}$ and simplify.

$$\frac{5a + 25}{2a} \cdot \frac{4a}{2a + 10}$$

Factor first.

$$\frac{5(a + 5)}{2a} \cdot \frac{(2)(2)a}{2(a + 5)}$$

Now cancel.

$$\frac{5\cancel{(a + 5)}}{\cancel{2a}} \cdot \frac{\cancel{(2)}\cancel{(2)}\cancel{a}}{\cancel{2}\cancel{(a + 5)}} = \frac{5}{1}$$

So $\dfrac{5a + 25}{2a} \cdot \dfrac{4a}{2a + 10} = 5.$

EXAMPLE 4   Simplify $\dfrac{x - 2y}{4x - 12} \cdot \dfrac{x^2 - 9}{2x - 4y} \cdot \dfrac{x}{2x + 6}$.

Factor first.

$$\frac{x - 2y}{4(x - 3)} \cdot \frac{(x + 3)(x - 3)}{2(x - 2y)} \cdot \frac{x}{2(x + 3)}$$

Now cancel.

$$\frac{\cancel{(x - 2y)}}{4\cancel{(x - 3)}} \cdot \frac{\cancel{(x + 3)}\cancel{(x - 3)}}{2\cancel{(x - 2y)}} \cdot \frac{x}{2\cancel{(x + 3)}} = \frac{x}{16}$$

So $\dfrac{x - 2y}{4x - 12} \cdot \dfrac{x^2 - 9}{2x - 4y} \cdot \dfrac{x}{2x + 6} = \dfrac{x}{16}$.

EXAMPLE 5   Simplify $\left(\dfrac{x^2 - 2x - 3}{x^2 - 9}\right)\left(\dfrac{x^2 + 5x + 6}{x^2 - 1}\right)$.

Factor first.

$$\frac{(x - 3)(x + 1)}{(x - 3)(x + 3)} \cdot \frac{(x + 3)(x + 2)}{(x - 1)(x + 1)}$$

Now cancel.

$$\frac{\overset{1}{\cancel{(x - 3)}}\overset{1}{\cancel{(x + 1)}}}{\underset{1}{\cancel{(x - 3)}}\underset{1}{\cancel{(x + 3)}}} \cdot \frac{\overset{1}{\cancel{(x + 3)}}(x + 2)}{(x - 1)\underset{1}{\cancel{(x + 1)}}} = \frac{x + 2}{x - 1}$$

So $\left(\dfrac{x^2 - 2x - 3}{x^2 - 9}\right)\left(\dfrac{x^2 + 5x + 6}{x^2 - 1}\right) = \dfrac{x + 2}{x - 1}$.

EXAMPLE 6   Simplify $\left(\dfrac{x^2 - y^2}{xy}\right)\left(\dfrac{y^2}{x + y}\right)$.

Factor first.

$$\frac{(x - y)(x + y)}{xy} \cdot \frac{yy}{(x + y)}$$

Now cancel.

$$\frac{(x - y)\overset{1}{\cancel{(x + y)}}}{\underset{1}{\cancel{xy}}} \cdot \frac{\overset{1}{y}\cancel{y}}{\underset{1}{\cancel{(x + y)}}} = \frac{y(x - y)}{x} \; \cancel{\;}$$

So $\left(\dfrac{x^2 - y^2}{xy}\right)\left(\dfrac{y^2}{x + y}\right) = \dfrac{y(x - y)}{x}$.

EXAMPLE 7 Simplify $\left(x^2 - 4\right)\left(\dfrac{5}{2 - x}\right)$.

Factor first.

$$\frac{(x + 2)(x - 2)}{1} \cdot \frac{5}{(2 - x)} = \frac{5(x + 2)(x - 2)}{1(2 - x)}$$

We don't have any identical factors on top and bottom. But notice that $(x - 2)$ on top and $(2 - x)$ on bottom are opposites and any expression divided by its opposite is equal to -1.

$$\frac{x - 2}{2 - x} = -1$$

This gives us

$$\frac{5(x + 2)(x - 2)}{1(2 - x)} \text{ is } \frac{5(x + 2)(-1)}{1} \text{ or } -5(x + 2)$$

So $\left(x^2 - 4\right)\left(\dfrac{5}{2 - x}\right) = -5(x + 2)$.

Division of Algebraic Fractions

Now we will <u>divide</u> algebraic fractions. Recall how we divide fractions. The division problem $\dfrac{4}{5} \div \dfrac{7}{9}$ is rewritten as the multiplication problem

$\dfrac{4}{5} \cdot \dfrac{9}{7}$. So $\dfrac{4}{5} \div \dfrac{7}{9}$ is $\dfrac{4}{5} \cdot \dfrac{9}{7}$ or $\dfrac{36}{35}$.

EXAMPLE 8 Simplify $\dfrac{-10c}{3d} \div \dfrac{4c}{6d}$.

Division problem: $\dfrac{-10c}{3d} \div \dfrac{4c}{6d}$

becomes multiplication problem: $\dfrac{-10c}{3d} \cdot \dfrac{6d}{4c}$

Factor.

$$\frac{-(5)(2)c}{3d} \cdot \frac{(2)(3)d}{(2)(2)c}$$

Now cancel.

$$\frac{-(5)(\cancel{2})\cancel{c}}{\cancel{3}\cancel{d}} \cdot \frac{(\cancel{2})(\cancel{3})\cancel{d}}{(\cancel{2})(\cancel{2})\cancel{c}} = \frac{-5}{1} \quad \text{or} \quad -5$$

So $\quad \dfrac{-10c}{3d} \div \dfrac{4c}{6d} = -5$.

EXAMPLE 9 Simplify $\dfrac{\frac{a^2}{b}}{a^3}$.

$$\frac{a^2}{b} \div \frac{a^3}{1} = \frac{a^2}{b} \cdot \frac{1}{a^3}$$

This means $\quad \dfrac{a^2}{b} \div a^3$.

Division problem: $\quad \dfrac{a^2}{b} \div \dfrac{a^3}{1}$

becomes multiplication problem: $\quad \dfrac{a^2}{b} \cdot \dfrac{1}{a^3}$

Factor.

$$\frac{aa}{b} \cdot \frac{1}{aaa}$$

Now cancel.

$$\frac{\cancel{a}\cancel{a}}{b} \cdot \frac{1}{\cancel{a}\cancel{a}a} = \frac{1}{ab}$$

So $\quad \dfrac{\frac{a^2}{b}}{a^3} = \dfrac{1}{ab}$

EXAMPLE 10 Simplify $\dfrac{3xy + x}{y^2 - y} \div \dfrac{3y + 1}{y}$.

Division problem:

$$\dfrac{3xy + x}{y^2 - y} \div \dfrac{3y + 1}{y}$$

becomes multiplication problem:

$$\dfrac{3xy + x}{y^2 - y} \cdot \dfrac{y}{3y + 1}$$

Factor.

$$\dfrac{x(3y + 1)}{y(y - 1)} \cdot \dfrac{y}{3y + 1}$$

Now cancel.

$$\dfrac{x\cancel{(3y + 1)}}{\cancel{y}(y - 1)} \cdot \dfrac{\cancel{y}}{\cancel{(3y + 1)}} = \dfrac{x}{y - 1}$$

So $\dfrac{3xy + x}{y^2 - y} \div \dfrac{3y + 1}{y} = \dfrac{x}{y - 1}$.

EXAMPLE 11 Simplify $\dfrac{2ab + b^2}{8a + 12b} \div \dfrac{2a^2 + ab}{6a + 9b}$.

1. Division problem:

$$\dfrac{2ab + b^2}{8a + 12b} \div \dfrac{2a^2 + ab}{6a + 9b}$$

2. becomes multiplication problem:

$$\dfrac{2ab + b^2}{8a + 12b} \cdot \dfrac{6a + 9b}{2a^2 + ab}$$

3. Factor.

$$\dfrac{b(2a + b)}{4(2a + 3b)} \cdot \dfrac{3(2a + 3b)}{a(2a + b)}$$

4. Now cancel.

$$\dfrac{b\cancel{(2a + b)}}{4\cancel{(2a + 3b)}} \cdot \dfrac{3\cancel{(2a + 3b)}}{a\cancel{(2a + b)}} = \dfrac{3b}{4a}$$

So $\dfrac{2ab + b^2}{8a + 12b} \div \dfrac{2a^2 + ab}{6a + 9b} = \dfrac{3b}{4a}$.

EXAMPLE 12 Simplify $\dfrac{x^2 - x - 2}{x^2 - 4} \div \dfrac{x^2 + 4x + 3}{x^2 + 5x + 6}$.

Division problem:

$$\frac{x^2 - x - 2}{x^2 - 4} \div \frac{x^2 + 4x + 3}{x^2 + 5x + 6}$$

becomes multiplication problem:

$$\frac{x^2 - x - 2}{x^2 - 4} \cdot \frac{x^2 + 5x + 6}{x^2 + 4x + 3}$$

Factor.

$$\frac{(x + 1)(x - 2)}{(x + 2)(x - 2)} \cdot \frac{(x + 2)(x + 3)}{(x + 1)(x + 3)}$$

Now cancel.

$$\frac{\cancel{(x + 1)}\cancel{(x - 2)}}{\cancel{(x + 2)}\cancel{(x - 2)}} \cdot \frac{\cancel{(x + 2)}\cancel{(x + 3)}}{\cancel{(x + 1)}\cancel{(x + 3)}} = 1$$

So $\dfrac{x^2 - x - 2}{x^2 - 4} \div \dfrac{x^2 + 4x + 3}{x^2 + 5x + 6} = 1$.

Exercises

Multiply and simplify.

1. $\dfrac{1}{2} \cdot \dfrac{1}{6}$

2. $-\dfrac{2}{3} \cdot \dfrac{3}{8}$

3. $\dfrac{12}{7} \cdot \dfrac{7}{8} \cdot \dfrac{1}{9}$

4. $\dfrac{a}{b} \cdot \dfrac{b}{c} \cdot c$

5. $\dfrac{2x}{y} \cdot \dfrac{y}{2}$

6. $x^2 \cdot \dfrac{1}{x}$

7. $\dfrac{xy}{z} \cdot \dfrac{z^2}{x^2}$

8. $\dfrac{x}{10} \cdot 5$

9. $3x^2 \cdot \dfrac{4}{6x^3}$

10. $\dfrac{2x^2y}{3ab} \cdot \dfrac{9a^2}{10xy}$

11. $abcde \cdot \dfrac{1}{abcde}$

12. $\dfrac{2x + 4}{3} \cdot \dfrac{3}{x + 2}$

13. $\dfrac{6y - 8}{7} \cdot \dfrac{7y}{3y - 4} \cdot \dfrac{1}{y^2}$

14. $\dfrac{2x + 2y}{2x} \cdot \dfrac{z}{xz + yz}$

15. $\dfrac{2x + 3y}{3x - 6y} \cdot \dfrac{4x - 8y}{4x + 6y}$

16. $\dfrac{8a}{a^2 - 16} \cdot \dfrac{a + 4}{4}$

17. $\dfrac{x^2 - 1}{3} \cdot \dfrac{2}{x - 1}$

18. $(x - 1)^2 \cdot \dfrac{2}{x - 1}$

19. $\dfrac{x^2 - 5x + 6}{2x - 4} \cdot \dfrac{2}{x - 3}$

20. $\dfrac{x^2 - x - 12}{x^2} \cdot \dfrac{x}{x - 4}$

21. $\dfrac{x - xy}{y} \cdot \dfrac{y^2}{2x - 2y}$

22. $\dfrac{x^2 + 2x + 1}{2x + 2} \cdot \dfrac{6}{3x + 3}$

23. $\dfrac{x^2 + 7x + 10}{x^2 - 3x - 10} \cdot \dfrac{2x - 10}{2x + 10}$

24. $\dfrac{x^2 - 9}{3 - x} \cdot \dfrac{y}{x + 3}$

25. $\dfrac{x^2 - 5x - 14}{7 - x} \cdot 2$

26. $\dfrac{2x^2 + 3x + 1}{1 + x} \cdot \dfrac{x + 2}{4x + 2}$

27. $\dfrac{x^3 + 10x^2 + 25x}{(x - 5)^2} \cdot \dfrac{x - 5}{(x + 5)^2}$

Simplify.

28. $\dfrac{a}{b} \div \dfrac{a}{b}$

29. $\dfrac{\dfrac{2x}{3y}}{\dfrac{4x}{9y}}$

30. $\dfrac{x^2 y}{z} \div \dfrac{xy}{z}$

31. $\dfrac{x + 2}{y} \div \dfrac{2x + 4}{y^2}$

32. $\dfrac{x^2 + 3x + 2}{x - 1} \div \dfrac{x + 2}{(x - 1)^2}$

33. $\dfrac{x^2 - 36}{(x + 6)^2} \div \dfrac{6x + 36}{x^2 + 12x + 36}$

34. $\dfrac{2x^2 + 5x + 3}{2x^2 + x - 3} \div \dfrac{x^2 - 1}{(x - 1)^2}$

35. $\dfrac{4x^3 - 9x}{4x^2 + 4x - 3} \div \dfrac{4x^2 + 14x - 30}{2x^2 + 9x - 5}$

Answers to Exercises

1. $\dfrac{1}{2} \cdot \dfrac{1}{6} = \dfrac{(1)\,(1)}{(2)(6)} = \dfrac{1}{12}$

2. $-\dfrac{2}{3} \cdot \dfrac{3}{8} = -\dfrac{\cancel{(2)}\cancel{(3)}}{(3)(2)(4)} = -\dfrac{1}{4}$

3. $\dfrac{12}{7} \cdot \dfrac{7}{8} \cdot \dfrac{1}{9} = \dfrac{(3)\,(4)\,\cancel{(7)}\,(1)}{\cancel{(7)}(2)\cancel{(4)}\cancel{(3)}(3)} = \dfrac{1}{6}$

4. $\dfrac{a}{b} \cdot \dfrac{b}{c} \cdot c = \dfrac{a \cdot \cancel{b} \cdot \cancel{c}}{\cancel{b} \cdot \cancel{c}} = a$

5. $\dfrac{2x}{y} \cdot \dfrac{y}{2} = \dfrac{2xy}{y \cdot 2} = x$

6. $x^{2} \cdot \dfrac{1}{x} = \dfrac{x \cdot x}{x} = x$

7. $\dfrac{xy}{z} \cdot \dfrac{z^{2}}{x^{2}} = \dfrac{x \cdot y \cdot z \cdot z}{z \cdot x \cdot x} = \dfrac{yz}{x}$

8. $\dfrac{x}{10} \cdot 5 = \dfrac{x \cdot 5}{2 \cdot 5} = \dfrac{x}{2}$

9. $3x^{2} \cdot \dfrac{4}{6x^{3}} = \dfrac{3 \cdot x \cdot x \cdot 2 \cdot 2}{3 \cdot 2 \cdot x \cdot x \cdot x} = \dfrac{2}{x}$

(10.) $\dfrac{2x^2y}{3ab} \cdot \dfrac{9a^2}{10xy} = \dfrac{2 \cdot x \cdot x \cdot y \cdot 3 \cdot 3 \cdot a \cdot a}{3 \cdot a \cdot b \cdot 2 \cdot 5 \cdot x \cdot y} = \dfrac{3ax}{5b}$

(11.) $abcde \cdot \dfrac{1}{a \cdot b \cdot c \cdot d \cdot e} = \dfrac{a \cdot b \cdot c \cdot d \cdot e}{a \cdot b \cdot c \cdot d \cdot e} = \dfrac{1}{1} = 1$

(12.) $\dfrac{2x+4}{3} \cdot \dfrac{3}{x+2} = \dfrac{2(x+2)(3)}{3(x+2)} = \dfrac{2}{1} = 2$

(13.) $\dfrac{6y-8}{7} \cdot \dfrac{7y}{3y-4} \cdot \dfrac{1}{y^2} = \dfrac{2(3y-4)(7)(y)}{(7)(3y-4)(y)(y)} = \dfrac{2}{y}$

(14.) $\dfrac{2x+2y}{2x} \cdot \dfrac{z}{xz+yz} = \dfrac{2(x+y)(z)}{(2)(x)(z)(x+y)} = \dfrac{1}{x}$

(15.) $\dfrac{2x+3y}{3x-6y} \cdot \dfrac{4x-8y}{4x+6y} = \dfrac{(2x+3y)(2)(2)(x-2y)}{3(x-2y)(2)(2x+3y)} = \dfrac{2}{3}$

(16.) $\dfrac{8a}{a^2-16} \cdot \dfrac{a+4}{4} = \dfrac{(4)(2a)(a+4)}{(a+4)(a-4)(4)} = \dfrac{2a}{a-4}$

(17.) $\dfrac{x^2-1}{3} \cdot \dfrac{2}{x-1} = \dfrac{(x+1)(x-1)(2)}{(3)(x-1)} = \dfrac{2(x+1)}{3}$

(18.) $(x-1)^2 \cdot \dfrac{2}{x-1} = \dfrac{(x-1)(x-1)(2)}{(x-1)} = 2(x-1)$

(19.) $\dfrac{x^2-5x+6}{2x-4} \cdot \dfrac{2}{x-3} = \dfrac{(x-2)(x-3)(2)}{2(x-2)(x-3)} = \dfrac{1}{1} = 1$

(20.) $\dfrac{x^2-x-12}{x^2} \cdot \dfrac{x}{x-4} = \dfrac{(x-4)(x+3)(x)}{(x)(x)(x-4)} = \dfrac{x+3}{x}$

(21.) $\dfrac{x-xy}{y} \cdot \dfrac{y^2}{2x-2y} = \dfrac{(x)(1-y)(y)(y)}{(y)(2)(x-y)} = \dfrac{xy(1-y)}{2(x-y)}$

(22.) $\dfrac{x^2+2x+1}{2x+2} \cdot \dfrac{6}{3x+3} = \dfrac{(x+1)(x+1)(2)(3)}{2(x+1)(3)(x+1)} = \dfrac{1}{1} = 1$

(23.) $\dfrac{x^2+7x+10}{x^2-3x-10} \cdot \dfrac{2x-10}{2x+10} = \dfrac{(x+2)(x+5)(2)(x-5)}{(x-5)(x+2)(2)(x+5)} = \dfrac{1}{1} = 1$

(24.) $\dfrac{x^2-9}{3-x} \cdot \dfrac{y}{x+3} = \dfrac{(x+3)(x-3)(y)}{(3-x)(x+3)} = (-1)(y) = -y$

(25.) $\dfrac{x^2-5x-14}{7-x} \cdot 2 = \dfrac{(x-7)(x+2)(2)}{(7-x)} = (-1)(2)(x+2) = -2(x+2)$

(26.) $\dfrac{2x^2+3x+1}{1+x} \cdot \dfrac{x+2}{4x+2} = \dfrac{(2x+1)(x+1)(x+2)}{(1+x)(2)(2x+1)} = \dfrac{(x+2)}{2}$

(27.) $\dfrac{x^3+10x^2+25x}{(x-5)^2} \cdot \dfrac{(x-5)}{(x+5)^2} = \dfrac{x(x^2+10x+25)(x-5)}{(x-5)(x-5)(x+5)(x+5)} =$

$\dfrac{x(x+5)(x+5)(x-5)}{(x-5)(x-5)(x+5)(x+5)} = \dfrac{x}{x-5}$

28. $\frac{a}{b} \div \frac{a}{b} = \frac{a}{b} \cdot \frac{b}{a} = \frac{(a)(b)}{(b)(a)} = \frac{1}{1} = 1$

29. $\dfrac{\frac{2x}{3y}}{\frac{4x}{9y}}$ means $\frac{2x}{3y} \div \frac{4x}{9y} = \frac{2x}{3y} \cdot \frac{9y}{4x} = \frac{2 \cdot 3 \cdot 3}{3 \cdot 2 \cdot 2} = \frac{3}{2}$

30. $\frac{x^2 y}{z} \div \frac{xy}{z} = \frac{x^2 y}{z} \cdot \frac{z}{xy} = \frac{x \cdot x \cdot y \cdot z}{z \cdot x \cdot y} = x$

31. $\frac{x+2}{y} \div \frac{2x+4}{y^2} = \frac{x+2}{y} \cdot \frac{y^2}{2x+4} = \frac{(x+2)(y)(y)}{y(2)(x+2)} = \frac{y}{2}$

32. $\frac{x^2+3x+2}{x-1} \div \frac{x+2}{(x-1)^2} = \frac{x^2+3x+2}{x-1} \cdot \frac{(x-1)^2}{x+2} =$

$\frac{(x+2)(x+1)(x-1)(x-1)}{(x-1)(x+2)} = (x+1)(x-1) = x^2-1$

33. $\frac{x^2-36}{(x+6)^2} \div \frac{6x+36}{x^2+12x+36} = \frac{x^2-36}{(x+6)^2} \cdot \frac{x^2+12x+36}{6x+36} =$

$\frac{(x+6)(x-6)(x+6)(x+6)}{(x+6)(x+6)6(x+6)} = \frac{x-6}{6}$

34. $\frac{2x^2+5x+3}{2x^2+x-3} \div \frac{x^2-1}{(x-1)^2} = \frac{2x^2+5x+3}{2x^2+x-3} \cdot \frac{(x-1)^2}{x^2-1} =$

$\frac{(2x+3)(x+1)(x-1)(x-1)}{(2x+3)(x-1)(x+1)(x-1)} = 1$

(35.) $\dfrac{4x^3-9x}{4x^2+4x-3} \div \dfrac{4x^2+14x-30}{2x^2+9x-5} =$

$\dfrac{4x^3-9x}{4x^2+4x-3} \cdot \dfrac{2x^2+9x-5}{4x^2+14x-30} =$

$\dfrac{x(2x-3)(2x+3)(2x-1)(x+5)}{(2x-1)(2x+3)(2x-3)(x+5)(2)} = \dfrac{x}{2}$

Additional Exercises

1. $\dfrac{1}{3} \cdot \dfrac{1}{4}$

2. $-\dfrac{3}{5} \cdot \dfrac{5}{6}$

3. $\dfrac{3}{4} \cdot \dfrac{12}{15} \cdot \dfrac{5}{6}$

4. $\dfrac{1}{3} \cdot \dfrac{3y}{2}$

5. $\dfrac{3x}{y} \cdot \dfrac{y^2}{x^3} \cdot x$

6. $\dfrac{xy}{xz} \cdot \dfrac{z^2}{yz}$

7. $x^3 \cdot \dfrac{1}{x^2}$

8. $\dfrac{x}{6} \cdot 4$

9. $5x^3 \cdot \dfrac{3}{10x^2}$

10. $\dfrac{12x^2}{5y} \cdot \dfrac{10y^3}{3x^2y}$

11. $\dfrac{xyz}{5} \cdot \dfrac{10}{x^2y^2z^2}$

12. $\dfrac{3x+9}{4} \cdot \dfrac{4}{x+3}$

13. $\dfrac{3a-6b}{5b} \cdot \dfrac{5a}{a-2b}$

14. $\dfrac{xy+xz}{x} \cdot \dfrac{2y}{3y+3z} \cdot 3$

15. $\dfrac{x+2y}{xy^2} \cdot \dfrac{2x^2y}{2x+4y}$

16. $\dfrac{a^2 - 4}{2b} \cdot \dfrac{4ab}{a + 2}$

17. $\dfrac{3}{x - 2} \cdot (x - 2)^2$

18. $\dfrac{xy + x}{x^2} \cdot \dfrac{x}{2y + 2}$

19. $\dfrac{xz + yz}{3z} \cdot \dfrac{x + y}{x^2 - y^2}$

20. $\dfrac{2xy + y^2}{3x + 9y} \cdot \dfrac{2x + 6y}{2x^2 + xy}$

21. $\dfrac{x^2 + 3x + 2}{3x + 3} \cdot \dfrac{3}{2x + 4}$

22. $\dfrac{x^2 + 6x + 8}{x^2 - x - 6} \cdot \dfrac{x^2 - 9}{x + 4}$

23. $\dfrac{x^2 - 4}{x^2 + 5x + 6} \cdot \dfrac{x^2 + 4x + 3}{x^2 - 1}$

24. $\dfrac{x^2 - 16}{4 - x} \cdot \dfrac{y}{x + 4}$

25. $\dfrac{x^2 - 7x + 10}{5 - x} \cdot 3$

26. $\dfrac{3x^2 + 7x + 2}{3x + 1} \cdot \dfrac{3x + 2}{2 + x}$

27. $\dfrac{x^3 - x}{(x - 1)^2} \cdot \dfrac{x - 1}{(x + 1)^2}$

Simplify.

28. $\dfrac{x}{y} \div \dfrac{z}{y}$

29. $a^2 \div a$

30. $\dfrac{\dfrac{2a}{3b}}{\dfrac{4a}{6b}}$

31. $\dfrac{a^2x}{b} \div \dfrac{ax}{b^2}$

32. $\dfrac{a^3}{\dfrac{a^2}{b}}$

33. $\dfrac{x + 3}{y^2} \div \dfrac{4x + 12}{y}$

34. $\dfrac{x^2 + 2x + 1}{x + 1} \div \dfrac{x + 2}{2x + 4}$

35. $\dfrac{x + 2y}{4xy - 12y^2} \div \dfrac{3x^3 + 6x^2y}{2xy - 6y^2}$

Addition and Subtraction of Algebraic Fractions

In this lesson and the next one, we will add and subtract algebraic fractions.

Fractions with Identical Denominators

Recall how we add fractions whose denominators (bottoms) are identical.

$$\frac{5}{9} + \frac{2}{9} = \frac{5 + 2}{9} \quad \text{or} \quad \frac{7}{9}$$

$$\frac{1}{7} - \frac{6}{7} = \frac{1 - 6}{7} \quad \text{or} \quad \frac{-5}{7}$$

We deal in the same way with algebraic fractions whose denominators are identical. But <u>remember</u>—the denominators can never be zero, since division by zero has no meaning.

EXAMPLE 1   Combine into a single fraction: $\dfrac{5}{x} + \dfrac{2}{x}$

$$\dfrac{5}{x} + \dfrac{2}{x}$$

$$= \dfrac{5 + 2}{x} \quad \text{or} \quad \dfrac{7}{x}$$

EXAMPLE 2   Combine into a single fraction: $\dfrac{1}{y} - \dfrac{6}{y}$

$$\dfrac{1}{y} - \dfrac{6}{y}$$

$$= \dfrac{1 - 6}{y} \quad \text{or} \quad \dfrac{-5}{y}$$

EXAMPLE 3   Combine into a single fraction: $\dfrac{2}{5z} - \dfrac{1}{5z} + \dfrac{6}{5z}$

$$\dfrac{2}{5z} - \dfrac{1}{5z} + \dfrac{6}{5z}$$

$$= \dfrac{2 - 1 + 6}{5z} \quad \text{or} \quad \dfrac{7}{5z}$$

EXAMPLE 4   Combine into a single fraction: $\dfrac{x - 3}{x^2 y^2} + \dfrac{7}{x^2 y^2} - \dfrac{x}{x^2 y^2}$

$$\dfrac{x - 3}{x^2 y^2} + \dfrac{7}{x^2 y^2} - \dfrac{x}{x^2 y^2}$$

$$= \dfrac{x - 3 + 7 - x}{x^2 y^2} \quad \text{or} \quad \dfrac{4}{x^2 y^2}$$

EXAMPLE 5   Combine into a single fraction: $\dfrac{-4}{a - b} + \dfrac{a + 1}{a - b} + \dfrac{4}{a - b}$

$$\dfrac{-4}{a - b} + \dfrac{a + 1}{a - b} + \dfrac{4}{a - b}$$

$$= \dfrac{-4 + a + 1 + 4}{a - b} \quad \text{or} \quad \dfrac{a + 1}{a - b}$$

EXAMPLE 6   Combine into a single fraction: $\dfrac{x}{y} - \dfrac{x-1}{y}$

$$\dfrac{x}{y} - \dfrac{x-1}{y}$$

$$= \dfrac{x - (x-1)}{y}$$

which is $\dfrac{x - x + 1}{y}$ or $\dfrac{1}{y}$.

EXAMPLE 7   Combine into a single fraction: $\dfrac{2a+3}{a+1} - \dfrac{a+2}{a+1}$

$$\dfrac{2a+3}{a+1} - \dfrac{a+2}{a+1}$$

$$= \dfrac{(2a+3) - (a+2)}{a+1}$$

which is $\dfrac{2a+3-a-2}{a+1}$

or $\dfrac{a+1}{a+1}$ or 1 .

EXAMPLE 8   Combine into a single fraction: $\dfrac{3a}{a+b} - \dfrac{a}{a+b} + \dfrac{2b}{a+b}$

$$\dfrac{3a}{a+b} - \dfrac{a}{a+b} + \dfrac{2b}{a+b}$$

$$= \dfrac{3a - a + 2b}{a+b}$$

or $\dfrac{2a+2b}{a+b}$

Note: If we factor 2 out of the top, we can simplify further.

$$\dfrac{2a+2b}{a+b} = \dfrac{2(a+b)}{a+b}$$

Now cancel.

$$\frac{2\overset{1}{\cancel{(a+b)}}}{\underset{1}{\cancel{(a+b)}}} = \frac{2}{1} \quad \text{or} \quad 2$$

So $\quad \dfrac{3a}{a+b} - \dfrac{a}{a+b} + \dfrac{2b}{a+b} = 2$.

EXAMPLE 9   Combine into a single fraction: $\dfrac{2x}{x+y} - \dfrac{3x}{x+y}$

$$\frac{2x}{x+y} - \frac{3x}{x+y} = \frac{2x-3x}{x+y} \quad \text{or} \quad \frac{-x}{x+y}$$

This fraction cannot be simplified.

Fractions with Different Denominators

If fractions have different denominators, then we must change them into
fractions with identical denominators before we can add or subtract.
(Recall, we did not do this for multiplication or division of fractions.) To add
$\dfrac{1}{6}$ and $\dfrac{1}{4}$, for example, we need to find a new denominator which has a 6 and
a 4 in it as factors. It would save us work if we had the lowest such number.
This number is called the **Lowest Common Denominator** or **LCD**.

6 splits into (3)(2) so the LCD must have a (3)(2) in it
4 splits into (2)(2) so the LCD must have a (2)(2) in it

So we choose   (3)(2)(2)   or   12 as the LCD.

We could have chosen (3)(2)(2)(2) or 24, but (3)(2)(2) is lower, does the job,
and is easier to work with.

Now $\qquad\qquad \dfrac{1}{6} + \dfrac{1}{4}$

can be written as $\quad \dfrac{1}{(3)(2)} + \dfrac{1}{(2)(2)}$

We want each denominator to be 12 or (3)(2)(2).

There is a 2 missing in the first denominator, so we multiply

$$\frac{1}{(3)(2)} \quad \text{by} \quad \frac{2}{2} \quad \text{to get} \quad \frac{2}{(3)(2)(2)} \; .$$

Notice that $\dfrac{2}{2}$ is 1, so we have changed only the <u>form</u> of the fraction, not its <u>value</u>.

There is a 3 missing in the second denominator, so we multiply

$$\frac{1}{(2)(2)} \cdot \frac{3}{3} = \frac{3}{(2)(2)(3)}$$

Notice that $\dfrac{3}{3}$ is 1, so again we have changed only the <u>form</u> of the fraction, not its <u>value</u>.

$$\frac{1}{6} + \frac{1}{4} = \frac{1}{(3)(2)} + \frac{1}{(2)(2)}$$

which is the same as $\quad \dfrac{1}{(3)(2)} \cdot \dfrac{2}{2} + \dfrac{1}{(2)(2)} \cdot \dfrac{3}{3}$

which is $\quad\quad\quad\quad \dfrac{2}{(3)(2)(2)} + \dfrac{3}{(2)(2)(3)}$

which is $\quad\quad\quad\quad \dfrac{5}{(3)(2)(2)} \quad \text{or} \quad \dfrac{5}{12}$

So $\quad \dfrac{1}{6} + \dfrac{1}{4} = \dfrac{5}{12}$.

In a similar way we deal with algebraic fractions whose denominators are different. But remember, the denominators can never be zero, since division by zero has no meaning.

EXAMPLE 10   Combine $\dfrac{1}{6x} + \dfrac{1}{4x}$ into a single fraction.

Note: The denominators are different, so to add these fractions, we find the LCD.

$6x$ splits into $(3)(2)(x)$
$4x$ splits into $(2)(2)(x)$

The LCD is $(3)(2)(x)(2)$ or $12x$.

Since $12x$ has a $(3)(2)(x)$ or $6x$ in it, and since it has a $(2)(2)(x)$ or $4x$ in it, it does the job.

$$\frac{1}{6x} + \frac{1}{4x} = \frac{1}{(3)(2)(x)} + \frac{1}{(2)(2)(x)}$$

But we want each denominator to be $(3)(2)(2)(x)$ or $12x$. Since there is a 2 missing in the first denominator, we multiply the first fraction by $\dfrac{2}{2}$, and since there is a 3 missing in the second denominator, we multiply the second fraction by $\dfrac{3}{3}$.

$$\frac{1}{(3)(2)(x)} + \frac{1}{(2)(2)(x)}$$

is

$$\frac{1}{(3)(2)(x)} \cdot \frac{2}{2} + \frac{1}{(2)(2)(x)} \cdot \frac{3}{3}$$

which is the same as

$$\frac{2}{(3)(2)(2)(x)} + \frac{3}{(3)(2)(2)(x)}$$

which is

$$\frac{5}{(3)(2)(2)(x)}$$

or

$$\frac{5}{12x}$$

So $\dfrac{1}{6x} + \dfrac{1}{4x} = \dfrac{5}{12x}$.

EXAMPLE 11 Combine $\dfrac{3}{x} - \dfrac{5}{xy}$ into a single fraction.

Note: The denominators are different, so we find the LCD.

x is prime

xy splits into $(x)(y)$

The LCD is $(x)(y)$ or xy.

Since xy has an x in it and an xy in it, it is the LCD.

We want each denominator to be xy.

Since there is a y missing in the first denominator, we multiply the first fraction by $\dfrac{y}{y}$. Since there is nothing missing in the second denominator, we leave it alone.

$$\frac{3}{x} - \frac{5}{(x)(y)}$$

is

$$\frac{3}{x} \cdot \frac{y}{y} - \frac{5}{(x)(y)}$$

which is the same as

$$\frac{3y}{(x)(y)} - \frac{5}{(x)(y)}$$

or

$$\frac{3y - 5}{(x)(y)}$$

So $\dfrac{3}{x} - \dfrac{5}{xy} = \dfrac{3y - 5}{xy}$.

EXAMPLE 12   Combine $\dfrac{2}{x} + \dfrac{3}{x^2}$ into a single fraction.

Since the denominators are different, we find the LCD.

x is prime

x^2 splits into $(x)(x)$

$$x$$
$$|$$

The LCD is $(x)(x)$ or x^2.

$$x^2$$

Since x^2 has an x in it and an $(x)(x)$ in it, it is the LCD.

So $\dfrac{2}{x} + \dfrac{3}{x^2} = \dfrac{2}{x} + \dfrac{3}{(x)(x)}$.

We want each denominator to be $(x)(x)$. There is an x missing in the first denominator, so we multiply the first fraction by $\dfrac{x}{x}$. There is nothing missing in the second denominator, so we leave it alone.

$$\dfrac{2}{x} \cdot \dfrac{x}{x} + \dfrac{3}{(x)(x)}$$

which is $\dfrac{2x}{(x)(x)} + \dfrac{3}{(x)(x)}$

or $\dfrac{2x + 3}{(x)(x)}$

So $\dfrac{2}{x} + \dfrac{3}{x^2} = \dfrac{2x + 3}{x^2}$.

EXAMPLE 13   Combine $2x - \dfrac{3}{x}$ into a single fraction.

First we write $2x$ as a fraction.

$$\dfrac{2x}{1} - \dfrac{3}{x}$$

Since the denominators are different, we find the LCD. 1 cannot be split and x cannot be split, so the LCD is $(1)(x)$ or x.

We want each denominator to be x. There is an x missing in the first denominator, so we multiply the first fraction by $\dfrac{x}{x}$. There is nothing missing in the second denominator, so we leave it alone.

$$\frac{2x}{1} - \frac{3}{x}$$

is the same as $\quad \dfrac{2x}{1} \cdot \dfrac{x}{x} - \dfrac{3}{x}$

which is $\quad\quad \dfrac{2x^2}{x} - \dfrac{3}{x}$

or $\quad\quad\quad \dfrac{2x^2 - 3}{x}$

So $\quad 2x - \dfrac{3}{x} = \dfrac{2x^2 - 3}{x}$.

EXAMPLE 14   Combine $\dfrac{2}{x^2 y} - \dfrac{5}{xy^2}$ into a single fraction.

Since the denominators are different, we find the LCD.

$\quad\quad x^2 y$ splits into $(x)(x)(y)$
$\quad\quad xy^2$ splits into $(x)(y)(y)$
$\quad\quad$ the LCD is $(x)(x)(y)(y)$

So $\quad \dfrac{2}{x^2 y} - \dfrac{5}{xy^2} = \dfrac{2}{(x)(x)(y)} - \dfrac{5}{(x)(y)(y)}$.

We want each denominator to be $(x)(x)(y)(y)$. There is a y missing from the first denominator, so we multiply the first fraction by $\dfrac{y}{y}$.

There is an x missing from the second denominator, so we multiply the second fraction by $\dfrac{x}{x}$.

$$\frac{2}{(x)(x)(y)} \cdot \frac{y}{y} - \frac{5}{(x)(y)(y)} \cdot \frac{x}{x}$$

is the same as $\dfrac{2y}{(x)(x)(y)(y)} - \dfrac{5x}{(x)(x)(y)(y)}$

which is $\dfrac{2y - 5x}{(x)(x)(y)(y)}$

or $\dfrac{2y - 5x}{x^2y^2}$

So $\dfrac{2}{x^2y} - \dfrac{5}{xy^2} = \dfrac{2y - 5x}{x^2y^2}$.

EXAMPLE 15   Combine $\dfrac{1}{a} + \dfrac{1}{b} + \dfrac{1}{c}$ into a single fraction.

Since the denominators are different, we find the LCD.

a is prime
b is prime
c is prime

and the LCD is $(a)(b)(c)$

$$\frac{1}{a} + \frac{1}{b} + \frac{1}{c}$$

is the same as $\dfrac{1}{a} \cdot \dfrac{bc}{bc} + \dfrac{1}{b} \cdot \dfrac{ac}{ac} + \dfrac{1}{c} \cdot \dfrac{ab}{ab}$

which is $\dfrac{bc}{abc} + \dfrac{ac}{abc} + \dfrac{ab}{abc}$

or $\dfrac{bc + ac + ab}{abc}$

So $\dfrac{1}{a} + \dfrac{1}{b} + \dfrac{1}{c} = \dfrac{bc + ac + ab}{abc}$.

EXAMPLE 16 Combine $\dfrac{a}{bc} + \dfrac{b}{ac} + \dfrac{c}{ab}$ into a single fraction.

Since the denominators are different, we find the LCD.

> bc splits into $(b)(c)$
> ac splits into $(a)(c)$
> ab splits into $(a)(b)$

and the LCD is $(a)(b)(c)$

$$\frac{a}{bc} + \frac{b}{ac} + \frac{c}{ab}$$

is the same as $\quad \dfrac{a}{bc} \cdot \dfrac{a}{a} + \dfrac{b}{ac} \cdot \dfrac{b}{b} + \dfrac{c}{ab} \cdot \dfrac{c}{c}$

which is $\qquad \dfrac{a^2}{abc} + \dfrac{b^2}{abc} + \dfrac{c^2}{abc}$

or $\qquad \dfrac{a^2 + b^2 + c^2}{abc}$

So $\quad \dfrac{a}{bc} + \dfrac{b}{ac} + \dfrac{c}{ab} = \dfrac{a^2 + b^2 + c^2}{abc}$.

Exercises

Combine into a single fraction and simplify when possible.

1. $\dfrac{3}{7} + \dfrac{2}{7} =$

2. $\dfrac{2}{5} + \dfrac{1}{5} =$

3. $\dfrac{8}{9} - \dfrac{4}{9} =$

4. $\dfrac{4}{9} - \dfrac{5}{9} =$

5. $\dfrac{3}{x} + \dfrac{2}{x} =$

6. $\dfrac{5}{x} + \dfrac{3}{x} =$

7. $\dfrac{3}{x} - \dfrac{2}{x} =$

8. $\dfrac{4}{a} - \dfrac{9}{a} =$

9. $\dfrac{3}{x} + \dfrac{5}{x} + \dfrac{2}{x} =$

10. $\dfrac{2}{x} + \dfrac{6}{x} - \dfrac{3}{x} =$

11. $\dfrac{3}{y} - \dfrac{8}{y} + \dfrac{2}{y} =$

12. $\dfrac{3}{2x} + \dfrac{5}{2x} + \dfrac{5}{2x} + \dfrac{1}{2x} =$

13. $\dfrac{4}{3y} + \dfrac{7}{3y} - \dfrac{2}{3y} =$

14. $\dfrac{5}{xy} + \dfrac{x}{xy} + \dfrac{2}{xy} =$

15. $\dfrac{4a}{bc} - \dfrac{5a}{bc} + \dfrac{2a}{bc} =$

16. $\dfrac{3a}{a+b} + \dfrac{4a}{a+b} =$

17. $\dfrac{7x}{x+y} - \dfrac{2x}{x+y} + \dfrac{y}{x+y} =$

18. $\dfrac{a+4}{a+b} + \dfrac{a-4}{a+b} =$

19. $\dfrac{-5}{x-y} + \dfrac{x+1}{x-y} + \dfrac{5}{x-y} =$

20. $\dfrac{a}{b} + \dfrac{a+1}{b} =$

21. $\dfrac{x}{y} - \dfrac{x-2}{y} =$

22. $\dfrac{4}{x} - \dfrac{x+4}{x} =$

23. $\dfrac{2x+3}{x+1} + \dfrac{x+4}{x+1} =$

24. $\dfrac{3x+5}{x+3} - \dfrac{x+1}{x+3} =$

25. $\dfrac{1}{2x} + \dfrac{1}{3x} =$

26. $\dfrac{1}{2x} + \dfrac{1}{4x} =$

27. $\dfrac{1}{3y} - \dfrac{1}{4y} =$

28. $\dfrac{2}{x} + \dfrac{7}{xy} =$

29. $\dfrac{4}{a} - \dfrac{3}{ab} =$

30. $\dfrac{3}{x} + \dfrac{5}{x^2} =$

31. $\dfrac{5}{a^2} - \dfrac{2}{a} =$

32. $3x - \dfrac{4}{x} =$

33. $\dfrac{2}{a} - 5a =$

34. $\dfrac{3}{ab} - \dfrac{4}{bc} =$

35. $\dfrac{5}{a^2b} - \dfrac{3}{ab^2} =$

36. $\dfrac{1}{x} + \dfrac{1}{y} + \dfrac{1}{z} =$

37. $\dfrac{1}{ab} + \dfrac{1}{ac} + \dfrac{1}{bc} =$

38. $\dfrac{x + 3}{4x} + \dfrac{2x + 1}{6x} =$

39. $\dfrac{6}{ab} - \dfrac{a - 3}{a^2 b^2} =$

40. $\dfrac{1}{y} - \dfrac{x + 1}{xy} =$

41. $\dfrac{3a + 1}{5a} - \dfrac{4a - 3}{4a} =$

42. $\dfrac{2y + 3}{12y} - \dfrac{5}{y} - \dfrac{3y - 6}{8y} =$

Answers to Exercises

(1.) $\dfrac{3}{7} + \dfrac{2}{7} =$

$\dfrac{3+2}{7} =$

$\dfrac{5}{7}$

(5.) $\dfrac{3}{x} + \dfrac{2}{x} =$

$\dfrac{3+2}{x} =$

$\dfrac{5}{x}$

(9.) $\dfrac{3}{x} + \dfrac{5}{x} + \dfrac{2}{x} =$

$\dfrac{3+5+2}{x} =$

$\dfrac{10}{x}$

(2.) $\dfrac{2}{5} + \dfrac{1}{5} =$

$\dfrac{2+1}{5} =$

$\dfrac{3}{5}$

(6.) $\dfrac{5}{x} + \dfrac{3}{x} =$

$\dfrac{5+3}{x} =$

$\dfrac{8}{x}$

(10.) $\dfrac{2}{x} + \dfrac{6}{x} - \dfrac{3}{x} =$

$\dfrac{2+6-3}{x} =$

$\dfrac{5}{x}$

(3.) $\dfrac{8}{9} - \dfrac{4}{9} =$

$\dfrac{8-4}{9} =$

$\dfrac{4}{9}$

(7.) $\dfrac{3}{x} - \dfrac{2}{x} =$

$\dfrac{3-2}{x} =$

$\dfrac{1}{x}$

(11.) $\dfrac{3}{y} - \dfrac{8}{y} + \dfrac{2}{y} =$

$\dfrac{3-8+2}{y} =$

$\dfrac{-3}{y}$

(4.) $\dfrac{4}{9} - \dfrac{5}{9} =$

$\dfrac{4-5}{9} =$

$\dfrac{-1}{9}$

(8.) $\dfrac{4}{a} - \dfrac{9}{a} =$

$\dfrac{4-9}{a} =$

$\dfrac{-5}{a}$

(12.) $\dfrac{3}{2x} + \dfrac{5}{2x} + \dfrac{5}{2x} + \dfrac{1}{2x} =$

$\dfrac{3+5+5+1}{2x} =$

$\dfrac{14}{2x}$ or $\dfrac{7}{x}$

(13.) $\dfrac{4}{3y} + \dfrac{7}{3y} - \dfrac{2}{3y} =$

$\dfrac{4+7-2}{3y} =$

$\dfrac{9}{3y} \quad or \quad \dfrac{3}{y}$

(14.) $\dfrac{5}{xy} + \dfrac{x}{xy} + \dfrac{2}{xy} =$

$\dfrac{5+x+2}{xy} =$

$\dfrac{7+x}{xy}$

(15.) $\dfrac{4a}{bc} - \dfrac{5a}{bc} + \dfrac{2a}{bc} =$

$\dfrac{4a-5a+2a}{bc} =$

$\dfrac{a}{bc}$

(16.) $\dfrac{3a}{a+b} + \dfrac{4a}{a+b} =$

$\dfrac{3a+4a}{a+b} =$

$\dfrac{7a}{a+b}$

(17.) $\dfrac{7x}{x+y} - \dfrac{2x}{x+y} + \dfrac{y}{x+y} =$

$\dfrac{7x-2x+y}{x+y} =$

$\dfrac{5x+y}{x+y}$

(18.) $\dfrac{a+4}{a+b} + \dfrac{a-4}{a+b} =$

$\dfrac{a+4+a-4}{a+b} =$

$\dfrac{2a}{a+b}$

(19.) $\dfrac{-5}{x-y} + \dfrac{x+1}{x-y} + \dfrac{5}{x-y} =$

$\dfrac{-5+x+1+5}{x-y} =$

$\dfrac{x+1}{x-y}$

(20.) $\dfrac{a}{b} + \dfrac{a+1}{b} =$

$\dfrac{a+a+1}{b} =$

$\dfrac{2a+1}{b}$

21. $\dfrac{x}{y} - \dfrac{x-2}{y} =$

$\dfrac{x-x+2}{y} =$

$\dfrac{2}{y}$

22. $\dfrac{4}{x} - \dfrac{x+4}{x} =$

$\dfrac{4-x-4}{x} =$

$\dfrac{-x}{x}$ or -1

23. $\dfrac{2x+3}{x+1} + \dfrac{x+4}{x+1} =$

$\dfrac{2x+3+x+4}{x+1} =$

$\dfrac{3x+7}{x+1}$

24. $\dfrac{3x+5}{x+3} - \dfrac{x+1}{x+3} =$

$\dfrac{3x+5-x-1}{x+3}$

$\dfrac{2x+4}{x+3}$

25. $\dfrac{1}{2x} + \dfrac{1}{3x} =$

$\dfrac{1}{2x} \cdot \dfrac{3}{3} + \dfrac{1}{3x} \cdot \dfrac{2}{2} =$

$\dfrac{3}{(2)(3)(x)} + \dfrac{2}{(2)(3)(x)} =$

$\dfrac{3+2}{(2)(3)(x)} =$

$\dfrac{5}{6x}$

26. $\dfrac{1}{2x} + \dfrac{1}{4x} =$

$\dfrac{1}{(2)(x)} \cdot \dfrac{2}{2} + \dfrac{1}{(2)(2)(x)} =$

$\dfrac{2+1}{(2)(2)(x)} =$

$\dfrac{3}{4x}$

(27.) $\dfrac{1}{3y} - \dfrac{1}{4y} =$

$\dfrac{1}{(3)(y)} \cdot \dfrac{4}{4} - \dfrac{1}{(4)(y)} \cdot \dfrac{3}{3} =$

$\dfrac{4}{(3)(4)(y)} - \dfrac{3}{(3)(4)(y)} =$

$\dfrac{1}{12y}$

(30.) $\dfrac{3}{x} + \dfrac{5}{x^2} =$

$\dfrac{3}{x} \cdot \dfrac{x}{x} + \dfrac{5}{(x)(x)} =$

$\dfrac{3x}{(x)(x)} + \dfrac{5}{(x)(x)} =$

$\dfrac{3x+5}{x^2}$

(28.) $\dfrac{2}{x} + \dfrac{7}{xy} =$

$\dfrac{2}{x} \cdot \dfrac{y}{y} + \dfrac{7}{(x)(y)} =$

$\dfrac{2y}{xy} + \dfrac{7}{xy} =$

$\dfrac{2y+7}{xy}$

(31.) $\dfrac{5}{a^2} - \dfrac{2}{a} =$

$\dfrac{5}{(a)(a)} - \dfrac{2}{a} \cdot \dfrac{a}{a} =$

$\dfrac{5}{aa} - \dfrac{2a}{aa} =$

$\dfrac{5-2a}{a^2}$

(29.) $\dfrac{4}{a} - \dfrac{3}{ab} =$

$\dfrac{4}{a} \cdot \dfrac{b}{b} - \dfrac{3}{(a)(b)} =$

$\dfrac{4b}{ab} - \dfrac{3}{ab} =$

$\dfrac{4b-3}{ab}$

(32.) $3x - \dfrac{4}{x} =$

$3x \cdot \dfrac{x}{x} - \dfrac{4}{x} =$

$\dfrac{3xx}{x} - \dfrac{4}{x} =$

$\dfrac{3x^2-4}{x}$

(33.) $\dfrac{2}{a} - 5a =$

$\dfrac{2}{a} - 5a \cdot \dfrac{a}{a} =$

$\dfrac{2}{a} - \dfrac{5a^2}{a} =$

$\dfrac{2 - 5a^2}{a}$

(35.) $\dfrac{5}{a^2 b} - \dfrac{3}{ab^2} =$

$\dfrac{5}{a^2 b} \cdot \dfrac{b}{b} - \dfrac{3}{ab^2} \cdot \dfrac{a}{a} =$

$\dfrac{5b}{a^2 b^2} - \dfrac{3a}{a^2 b^2} =$

$\dfrac{5b - 3a}{a^2 b^2}$

(34.) $\dfrac{3}{ab} - \dfrac{4}{bc} =$

$\dfrac{3}{ab} \cdot \dfrac{c}{c} - \dfrac{4}{bc} \cdot \dfrac{a}{a} =$

$\dfrac{3c}{abc} - \dfrac{4a}{abc} =$

$\dfrac{3c - 4a}{abc}$

(36.) $\dfrac{1}{x} + \dfrac{1}{y} + \dfrac{1}{z} =$

$\dfrac{1}{x} \cdot \dfrac{yz}{yz} + \dfrac{1}{y} \cdot \dfrac{xz}{xz} + \dfrac{1}{z} \cdot \dfrac{xy}{xy} =$

$\dfrac{yz}{xyz} + \dfrac{xz}{xyz} + \dfrac{xy}{xyz} =$

$\dfrac{yz + xz + xy}{xyz}$

(37.) $\dfrac{1}{ab} + \dfrac{1}{ac} + \dfrac{1}{bc} =$

$\dfrac{1}{ab} \cdot \dfrac{c}{c} + \dfrac{1}{ac} \cdot \dfrac{b}{b} + \dfrac{1}{bc} \cdot \dfrac{a}{a} =$

$\dfrac{c}{abc} + \dfrac{b}{abc} + \dfrac{a}{abc} =$

$\dfrac{c + b + a}{abc} \quad \text{or} \quad \dfrac{a + b + c}{abc}$

(38.) $\dfrac{x+3}{4x} + \dfrac{2x+1}{6x} =$

$\dfrac{x+3}{(2)(2)(x)} \cdot \dfrac{3}{3} + \dfrac{2x+1}{(2)(3)(x)} \cdot \dfrac{2}{2} =$

$\dfrac{3(x+3)}{(2)(2)(3)(x)} + \dfrac{2(2x+1)}{(2)(2)(3)(x)} =$

$\dfrac{3x+9+4x+2}{12x} =$

$\dfrac{7x+11}{12x}$

(40.) $\dfrac{1}{y} - \dfrac{x+1}{xy} =$

$\dfrac{1}{y} \cdot \dfrac{x}{x} - \dfrac{x+1}{xy} =$

$\dfrac{x}{xy} - \dfrac{x+1}{xy} =$

$\dfrac{x-x-1}{xy} =$

$\dfrac{-1}{xy}$

(39.) $\dfrac{6}{ab} - \dfrac{a-3}{a^2b^2} =$

$\dfrac{6}{ab} \cdot \dfrac{ab}{ab} - \dfrac{a-3}{(a)(a)(b)(b)} =$

$\dfrac{6ab}{a \cdot a \cdot b \cdot b} - \dfrac{a-3}{a \cdot a \cdot b \cdot b} =$

$\dfrac{6ab-a+3}{a^2b^2}$

(41.) $\dfrac{3a+1}{5a} - \dfrac{4a-3}{4a} =$

$\dfrac{3a+1}{5 \cdot a} \cdot \dfrac{4}{4} - \dfrac{4a-3}{4 \cdot a} \cdot \dfrac{5}{5} =$

$\dfrac{4(3a+1)}{(5)(4)(a)} - \dfrac{5(4a-3)}{(4)(5)(a)} =$

$\dfrac{12a+4-20a+15}{20a}$

$\dfrac{-8a+19}{20a}$

42.) $\dfrac{2y+3}{12y} - \dfrac{5}{y} - \dfrac{3y-6}{8y} =$

$\dfrac{2y+3}{12y} \cdot \dfrac{2}{2} - \dfrac{5}{y} \cdot \dfrac{24}{24} - \dfrac{3y-6}{8y} \cdot \dfrac{3}{3} =$

$\dfrac{2(2y+3)}{24y} - \dfrac{24(5)}{24y} - \dfrac{3(3y-6)}{24y} =$

$\dfrac{4y+6-120-9y+18}{24y}$ or $\dfrac{-5y-96}{24y}$

Additional Exercises

Combine into a single fraction and simplify when possible.

1. $\dfrac{5}{3} + \dfrac{2}{3} =$

2. $\dfrac{5}{7} + \dfrac{3}{7} =$

3. $\dfrac{9}{5} + \dfrac{2}{5} =$

4. $\dfrac{2}{11} - \dfrac{4}{11} =$

5. $\dfrac{4}{5} - \dfrac{5}{5} =$

6. $\dfrac{2}{9} - \dfrac{5}{9} =$

7. $\dfrac{5}{y} + \dfrac{7}{y} =$

8. $\dfrac{3}{x} + \dfrac{2}{x} =$

9. $\dfrac{3}{x} + \dfrac{5}{x} =$

10. $\dfrac{3}{x} + \dfrac{1}{x} =$

11. $\dfrac{5}{y} - \dfrac{9}{y} =$

12. $\dfrac{7}{b} - \dfrac{1}{b} =$

13. $\dfrac{-2}{a} + \dfrac{5}{a} + \dfrac{3}{a} =$

14. $\dfrac{6}{7x} - \dfrac{2}{7x} + \dfrac{5}{7x} =$

15. $\dfrac{3}{x^2} - \dfrac{2}{x^2} + \dfrac{1}{x^2} =$

16. $\dfrac{3a}{bc} - \dfrac{a}{bc} + \dfrac{2a}{bc} =$

17. $\dfrac{7}{x + y} - \dfrac{2}{x + y} - \dfrac{7}{x + y} =$

18. $\dfrac{2a}{7c} - \dfrac{3b}{7c} + \dfrac{2a}{7c} =$

19. $\dfrac{2(x - 1)}{x^2 y^2} + \dfrac{3(x + 2)}{x^2 y^2} - \dfrac{5(x + 2)}{x^2 y^2} =$

20. $\dfrac{2a + b}{c} - \dfrac{a - b}{c} + \dfrac{3a}{c} =$

21. $\dfrac{5}{y} - \dfrac{a + 5}{y} =$

22. $\dfrac{3x + 4}{x + 3} + \dfrac{2x + 5}{x + 3} =$

23. $\dfrac{1}{5x} + \dfrac{1}{4x} =$

24. $\dfrac{1}{8y} + \dfrac{1}{12y} =$

25. $\dfrac{1}{2a} - \dfrac{1}{3a} =$

26. $\dfrac{3}{x} - \dfrac{5}{xy} =$

27. $\dfrac{5}{ab} - \dfrac{4}{b} =$

28. $\dfrac{4}{x} + \dfrac{7}{x^2} =$

29. $\dfrac{5}{x^3} - \dfrac{3}{x^2} + \dfrac{1}{x} =$

30. $5y - \dfrac{3}{y} =$

31. $\dfrac{5}{xy} - \dfrac{2}{yz} =$

32. $\dfrac{4}{x^2 y} - \dfrac{2}{xy^2} - \dfrac{5}{xy} =$

33. $\dfrac{1}{a} + \dfrac{2}{b} + \dfrac{3}{c} =$

34. $\dfrac{5}{xy} - \dfrac{4}{yz} - \dfrac{3}{xz} =$

35. $\dfrac{x+5}{3x} + \dfrac{3x-2}{5x} =$

36. $\dfrac{5}{xy} - \dfrac{x-2}{x^2y^2} =$

37. $\dfrac{1}{a} - \dfrac{a+3}{ab} =$

38. $\dfrac{2a+3}{3a} - \dfrac{5a-2}{4a} =$

39. $\dfrac{x-3}{2x} - \dfrac{2x+5}{x} + \dfrac{5}{3x} =$

40. $\dfrac{3a}{2} - \dfrac{2a-3}{20a} + \dfrac{3a-4}{5a} =$

Advanced Addition and Subtraction of Algebraic Fractions

In this lesson we will combine fractions with more complicated denominators.

EXAMPLE 1   Combine $\dfrac{x}{x+1} + \dfrac{1}{y}$ into a single fraction.

Note: The denominators are different, so to add these fractions we find the LCD.

> $x + 1$ is prime.
> y is prime

and the LCD is $(x + 1)(y)$.

So we want each denominator to be $(x + 1)(y)$. There is a y missing in the first denominator, so we multiply the first fraction by $\dfrac{y}{y}$. There is an $x + 1$ missing in the second denominator, so we multiply the second fraction by $\dfrac{x + 1}{x + 1}$.

$$\frac{x}{(x+1)} \cdot \frac{y}{y} + \frac{1}{y} \cdot \frac{x+1}{x+1}$$

is the same as $\dfrac{xy}{(x+1)(y)} + \dfrac{x+1}{y(x+1)}$

which is $\dfrac{xy + x + 1}{y(x+1)}$

So $\dfrac{x}{x+1} + \dfrac{1}{y} = \dfrac{xy + x + 1}{y(x+1)}$.

EXAMPLE 2   Combine $\dfrac{1}{x} + \dfrac{1}{x+3}$ into a single fraction.

Notice that the denominators are different. So to add these fractions, we find the LCD.

x is prime
$x + 3$ is prime

and the LCD is $(x)(x+3)$.

So we want each denominator to be $(x)(x+3)$. There is an $x + 3$ missing in the first denominator, so we multiply the first fraction by $\dfrac{x+3}{x+3}$. There is an x missing in the second denominator, so we multiply the second fraction by $\dfrac{x}{x}$.

$$\frac{1}{x} \cdot \frac{x+3}{x+3} + \frac{1}{x+3} \cdot \frac{x}{x}$$

is the same as $\dfrac{x+3}{x(x+3)} + \dfrac{x}{(x+3)x}$

which is $\dfrac{x+3+x}{x(x+3)}$ or $\dfrac{2x+3}{x(x+3)}$

So $\dfrac{1}{x} + \dfrac{1}{x+3} = \dfrac{2x+3}{x(x+3)}$.

EXAMPLE 3   Combine $\dfrac{1}{x+1} + \dfrac{1}{x-1}$ into a single fraction.

Find the LCD.

$x + 1$ is prime
$x - 1$ is prime

and the LCD is $(x+1)(x-1)$.

So $\dfrac{1}{x+1} + \dfrac{1}{x-1}$

is the same as $\dfrac{1}{x+1} \cdot \dfrac{x-1}{x-1} + \dfrac{1}{x-1} \cdot \dfrac{x+1}{x+1}$

which is $\dfrac{x-1}{(x+1)(x-1)} + \dfrac{x+1}{(x-1)(x+1)}$

or $\dfrac{x-1+x+1}{(x+1)(x-1)}$ or $\dfrac{2x}{(x+1)(x-1)}$

So $\dfrac{1}{x+1} + \dfrac{1}{x-1} = \dfrac{2x}{(x+1)(x-1)}$.

EXAMPLE 4   Combine $5 - \dfrac{1}{x+y}$ into a single fraction.

First write 5 as a fraction.

$\dfrac{5}{1} \quad \dfrac{1}{x+y}$

Now find the LCD.

$x + y$ is prime

and the LCD is $(1)(x+y)$ or $x + y$.

So $5 - \dfrac{1}{x+y}$

is the same as $\dfrac{5}{1} \cdot \dfrac{x+y}{x+y} - \dfrac{1}{x+y}$

which is
$$\frac{5(x + y)}{x + y} - \frac{1}{x + y}$$

or
$$\frac{5(x + y) - 1}{x + y}$$

Now we simplify by removing the parentheses on the top.

$$\frac{5x + 5y - 1}{x + y}$$

So
$$5 - \frac{1}{x + y} = \frac{5x + 5y - 1}{x + y}.$$

EXAMPLE 5   Combine $\dfrac{2x - 3}{2} - \dfrac{3x - 2}{3}$ into a single fraction.

Find the LCD.

    2 is prime
    3 is prime

and the LCD is (2)(3).

So
$$\frac{2x - 3}{2} - \frac{3x - 2}{3}$$

is the same as
$$\left(\frac{2x - 3}{2}\right)\frac{3}{3} - \left(\frac{3x - 2}{3}\right)\frac{2}{2}$$

which is
$$\frac{6x - 9}{6} - \frac{6x - 4}{6}$$

or
$$\frac{(6x - 9) - (6x - 4)}{6}$$

Now we simplify by removing the parentheses on the top.

$$\frac{6x - 9 - 6x + 4}{6} = \frac{-5}{6}$$

So
$$\frac{2x - 3}{2} - \frac{3x - 2}{3} = -\frac{5}{6}.$$

EXAMPLE 6 Combine $\dfrac{2}{3} - \dfrac{5}{3x - 9}$ into a single fraction.

To find the LCD, we factor each denominator.

3 is prime
$3x - 9$ splits into $3(x - 3)$

and the LCD is $3(x - 3)$.

We rewrite the problem with the factored denominators.

$$\frac{2}{3} - \frac{5}{3(x - 3)}$$

is the same as $\dfrac{2}{3} \cdot \dfrac{x - 3}{x - 3} - \dfrac{5}{3(x - 3)}$

which is $\dfrac{2(x - 3)}{3(x - 3)} - \dfrac{5}{3(x - 3)}$

or $\dfrac{2(x - 3) - 5}{3(x - 3)}$

Now we simplify by removing the parentheses on the top.

$$\frac{2x - 6 - 5}{3(x - 3)} = \frac{2x - 11}{3(x - 3)}$$

So $\dfrac{2}{3} - \dfrac{5}{3x - 9} = \dfrac{2x - 11}{3(x - 3)}$.

EXAMPLE 7 Combine $\dfrac{4x}{x^2 - 1} + \dfrac{5}{x - 1}$ into a single fraction.

To find the LCD, we factor each denominator.

$x^2 - 1$ splits into $(x - 1)(x + 1)$ $x^2 - 1$
$x - 1$ is prime

and the LCD is $(x - 1)(x + 1)$.

We rewrite the problem with factored denominators.

$$\frac{4x}{(x-1)(x+1)} + \frac{5}{x-1}$$

is the same as $\dfrac{4x}{(x-1)(x+1)} + \dfrac{5}{x-1} \cdot \dfrac{x+1}{x+1}$

which is $\dfrac{4x}{(x-1)(x+1)} + \dfrac{5(x+1)}{(x-1)(x+1)}$

which is $\dfrac{4x + 5(x+1)}{(x-1)(x+1)}$

Now we simplify by removing the parentheses on the top.

$$\frac{4x + 5x + 5}{(x-1)(x+1)} = \frac{9x+5}{(x-1)(x+1)}$$

So $\dfrac{4x}{x^2-1} + \dfrac{5}{x-1} = \dfrac{9x+5}{(x-1)(x+1)}$.

EXAMPLE 8 Combine $\dfrac{5}{x^2-4} - \dfrac{4}{x-2} + \dfrac{3}{2+x}$ into a single fraction.

To find the LCD, we factor each denominator.

$x^2 - 4$ splits into $(x+2)(x-2)$
$x - 2$ is prime
$2 + x$ is the same as $x + 2$ and is prime

and the LCD is $(x+2)(x-2)$.

We rewrite the problem with factored denominators.

$$\frac{5}{(x+2)(x-2)} - \frac{4}{x-2} + \frac{3}{2+x}$$

is the same as $\dfrac{5}{(x+2)(x-2)} - \dfrac{4}{x-2} \cdot \dfrac{x+2}{x+2} + \dfrac{3}{2+x} \cdot \dfrac{x-2}{x-2}$

which is $\dfrac{5}{(x+2)(x-2)} - \dfrac{4(x+2)}{(x-2)(x+2)} + \dfrac{3(x-2)}{(2+x)(x-2)}$

which is $\dfrac{5 - 4(x+2) + 3(x-2)}{(x+2)(x-2)}$

Now we simplify by removing the parentheses on the top.

$$\frac{5 - 4x - 8 + 3x - 6}{(x + 2)(x - 2)} = \frac{-x - 9}{(x + 2)(x - 2)}$$

So $\quad \dfrac{5}{x^2 - 4} - \dfrac{4}{x - 2} + \dfrac{3}{2 + x} = \dfrac{-x - 9}{(x + 2)(x - 2)} \, .$

EXAMPLE 9 Combine $\quad \dfrac{2x}{3x + 12} + \dfrac{3}{2x + 8}$ into a single fraction.

To find the LCD, we factor each denominator.

$\quad 3x + 12$ splits into $3(x + 4)$
$\quad 2x + 8$ splits into $2(x + 4)$

and the LCD is $(3)(2)(x + 4)$.

We rewrite the problem with factored denominators.

$$\frac{2x}{3(x + 4)} + \frac{3}{2(x + 4)}$$

is the same as $\quad \dfrac{2x}{3(x + 4)} \cdot \dfrac{2}{2} + \dfrac{3}{2(x + 4)} \cdot \dfrac{3}{3}$

which is $\quad \dfrac{4x}{(3)(x + 4)(2)} + \dfrac{9}{2(x + 4)(3)}$

or $\quad \dfrac{4x + 9}{(3)(2)(x + 4)}$

So $\quad \dfrac{2x}{3x + 12} + \dfrac{3}{2x + 8} = \dfrac{4x + 9}{6(x + 4)} \, .$

EXAMPLE 10 Combine $\dfrac{1}{x-5} - \dfrac{x+4}{x^2-10x+25}$ into a single fraction.

To find the LCD, we factor each denominator.

$x-5$ is prime

$x^2 - 10x + 25$ splits into $(x-5)(x-5)$

and the LCD is $(x-5)(x-5)$.

We rewrite the problem with the factored denominator.

$$\dfrac{1}{x-5} - \dfrac{x+4}{(x-5)(x-5)}$$

is the same as $\dfrac{1}{x-5} \cdot \dfrac{x-5}{x-5} - \dfrac{x+4}{(x-5)(x-5)}$

which is $\dfrac{x-5}{(x-5)(x-5)} - \dfrac{x+4}{(x-5)(x-5)}$

or $\dfrac{(x-5)-(x+4)}{(x-5)(x-5)}$

Now we simplify by removing the parentheses on the top.

$$= \dfrac{x-5-x-4}{(x-5)(x-5)} \qquad \dfrac{-9}{(x-5)(x-5)}$$

why (-4) *sign Change*

So $\dfrac{1}{x-5} - \dfrac{x+4}{x^2-10x+25} = \dfrac{-9}{(x-5)(x-5)}$.

EXAMPLE 11 Combine $\dfrac{5}{x^2 - 9} + \dfrac{2}{x^2 - 6x + 9}$ into a single fraction.

To find the LCD, we factor each denominator.

$x^2 - 9$ splits into $(x + 3)(x - 3)$
$x^2 - 6x + 9$ splits into $(x - 3)(x - 3)$

$$x^2 - 9$$

and the LCD is $(x + 3)(x - 3)(x - 3)$.

$$x^2 - 6x + 9$$

We rewrite the problem with factored denominators.

$$\frac{5}{(x + 3)(x - 3)} + \frac{2}{(x - 3)(x - 3)}$$

is the same as $\dfrac{5}{(x + 3)(x - 3)} \cdot \dfrac{x - 3}{x - 3} + \dfrac{2}{(x - 3)(x - 3)} \cdot \dfrac{x + 3}{x + 3}$

which is $\dfrac{5(x - 3)}{(x + 3)(x - 3)(x - 3)} + \dfrac{2(x + 3)}{(x - 3)(x - 3)(x + 3)}$

or $\dfrac{5(x - 3) + 2(x + 3)}{(x + 3)(x - 3)(x - 3)}$

Now we simplify by removing the parentheses on top.

$$\frac{5x - 15 + 2x + 6}{(x + 3)(x - 3)(x - 3)} = \frac{7x - 9}{(x + 3)(x - 3)^2}$$

So $\dfrac{5}{x^2 - 9} + \dfrac{2}{x^2 - 6x + 9} = \dfrac{7x - 9}{(x + 3)(x - 3)^2}$.

Exercises

Combine into a single fraction and simplify.

1. $\dfrac{a}{a + 3} + \dfrac{1}{2}$

2. $\dfrac{3}{y} + \dfrac{2}{y + 1}$

3. $\dfrac{2x}{x - 2} - \dfrac{2}{x}$

4. $\dfrac{1}{x + 3} + \dfrac{1}{x - 3}$

5. $\dfrac{5a}{2a + 4} - \dfrac{5}{6}$

6. $3 + \dfrac{2}{a} + \dfrac{4}{b}$

7. $\dfrac{a}{c} + \dfrac{a}{b} - \dfrac{5}{bc}$

8. $\dfrac{x}{x + y} + \dfrac{y}{x - y}$

9. $\dfrac{5}{x + 3} + \dfrac{3}{x - 3} - \dfrac{2}{x^2 - 9}$

10. $\dfrac{3}{x - 1} - \dfrac{4}{x^2 - 1} + 2$

11. $\dfrac{3a}{2a + 6} - \dfrac{4a}{3a + 9}$

12. $\dfrac{5}{3x - 15} + \dfrac{2}{5x - 25}$

13. $\dfrac{7a}{a^2 + 2a - 3} + \dfrac{2a}{a^2 + 5a + 6} - \dfrac{3a}{a^2 + a - 2}$

14. $\dfrac{x}{x^2 - 9} + \dfrac{5}{x + 3} - 3$

15. $\dfrac{a}{a - 2} + \dfrac{3(2a - 9)}{a^2 - 5a + 6}$

16. $\dfrac{y-2}{4y} - \dfrac{3y+5}{3y}$

17. $\dfrac{3a-b}{2a-b} + \dfrac{3a^2}{4a^2-b^2}$

18. $\dfrac{1}{a+b} + \dfrac{1}{b+a} \quad = \dfrac{1}{a+b}$

19. $\dfrac{a-b}{a+b} + \dfrac{a+b}{a+b} \quad = \dfrac{2a}{a+b}$

20. $\dfrac{6}{x^3+4x} + \dfrac{3x}{x^2+4} - \dfrac{2}{x}$

Answers to Exercises

$$\textcircled{1.}\ \frac{a}{a+3} + \frac{1}{2} =$$

$$\frac{a}{a+3} \cdot \frac{2}{2} + \frac{1}{2} \cdot \frac{a+3}{a+3} =$$

$$\frac{2a}{2(a+3)} + \frac{1(a+3)}{2(a+3)} =$$

$$\frac{2a+1a+3}{2(a+3)} =$$

$$\frac{3a+3}{2(a+3)}\ or\ \frac{3(a+1)}{2(a+3)}$$

$$\textcircled{3.}\ \frac{2x}{x-2} - \frac{2}{x} =$$

$$\frac{2x}{(x-2)} \cdot \frac{x}{x} - \frac{2}{x} \cdot \frac{(x-2)}{(x-2)} =$$

$$\frac{2x^2}{x(x-2)} - \frac{2(x-2)}{x(x-2)} =$$

$$\frac{2x^2-2(x-2)}{x(x-2)}\ or$$

$$\frac{2x^2-2x+4}{x(x-2)}$$

$$\textcircled{2.}\ \frac{3}{y} + \frac{2}{y+1} =$$

$$\frac{3}{y} \cdot \frac{(y+1)}{(y+1)} + \frac{2}{(y+1)} \cdot \frac{y}{y} =$$

$$\frac{3(y+1)}{y(y+1)} + \frac{2y}{y(y+1)} =$$

$$\frac{3y+3}{y(y+1)} + \frac{2y}{y(y+1)} =$$

$$\frac{3y+3+2y}{y(y+1)}\ or\ \frac{5y+3}{y(y+1)}$$

$$\textcircled{4.}\ \frac{1}{x+3} + \frac{1}{x-3} =$$

$$\frac{1}{(x+3)} \cdot \frac{(x-3)}{(x-3)} + \frac{1}{(x-3)} \cdot \frac{(x+3)}{(x+3)} =$$

$$\frac{1(x-3)}{(x+3)(x-3)} + \frac{1(x+3)}{(x+3)(x+3)} =$$

$$\frac{1x-3+1x+3}{(x+3)(x-3)}\ or$$

$$\frac{2x}{(x+3)(x-3)}$$

(5.) $\dfrac{5a}{2a+4} - \dfrac{5}{6}$

$\dfrac{5a}{2(a+2)} \cdot \dfrac{3}{3} - \dfrac{5}{6} \cdot \dfrac{(a+2)}{(a+2)} =$

$\dfrac{3(5a)}{2(a+2)(3)} - \dfrac{5(a+2)}{6(a+2)} =$

$\dfrac{3(5a) - 5(a+2)}{6(a+2)} =$

$\dfrac{15a - 5a - 10}{6(a+2)}$ or $\dfrac{10(a-1)}{6(a+2)}$

which is $\dfrac{5(a-1)}{3(a+2)}$

(6.) $3 + \dfrac{2}{a} + \dfrac{4}{b} =$

$\dfrac{3}{1} \cdot \dfrac{ab}{ab} + \dfrac{2}{a} \cdot \dfrac{b}{b} + \dfrac{4}{b} \cdot \dfrac{a}{a} =$

$\dfrac{3(ab)}{ab} + \dfrac{2b}{ab} + \dfrac{4a}{ab} =$

$\dfrac{3ab + 2b + 4a}{ab}$

(7.) $\dfrac{a}{c} + \dfrac{a}{b} - \dfrac{5}{bc} =$

$\dfrac{a}{c} \cdot \dfrac{b}{b} + \dfrac{a}{b} \cdot \dfrac{c}{c} - \dfrac{5}{bc} =$

$\dfrac{ab}{cb} + \dfrac{ac}{bc} - \dfrac{5}{bc} =$

$\dfrac{ab + ac - 5}{bc}$

(8.) $\dfrac{x}{x+y} + \dfrac{y}{x-y} =$

$\dfrac{x}{(x+y)} \cdot \dfrac{(x-y)}{(x-y)} + \dfrac{y}{(x-y)} \cdot \dfrac{(x+y)}{(x+y)} =$

$\dfrac{x(x-y)}{(x+y)(x-y)} + \dfrac{y(x+y)}{(x+y)(x-y)} =$

$\dfrac{x(x-y) + y(x+y)}{(x+y)(x-y)} =$

$\dfrac{x^2 + y^2}{x^2 - y^2}$

(9.) $\dfrac{5}{X+3} + \dfrac{3}{X-3} - \dfrac{2}{X^2-9} =$

$\dfrac{5}{X+3} \cdot \dfrac{X-3}{X-3} + \dfrac{3}{X-3} \cdot \dfrac{X+3}{X+3} - \dfrac{2}{(X+3)(X-3)} =$

$\dfrac{5(X-3)+3(X+3)-2}{(X+3)(X-3)} =$

$\dfrac{5X-15+3X+9-2}{(X+3)(X-3)} =$

$\dfrac{8X-8}{(X+3)(X-3)}$ or $\dfrac{8(X-1)}{(X+3)(X-3)}$

(10.) $\dfrac{3}{X-1} - \dfrac{4}{X^2-1} + 2 =$

$\dfrac{3}{(X-1)} \cdot \dfrac{(X+1)}{(X+1)} - \dfrac{4}{(X+1)(X-1)} + 2 \cdot \dfrac{(X+1)(X-1)}{(X+1)(X-1)} =$

$\dfrac{3(X+1)-4+2(X+1)(X-1)}{(X+1)(X-1)} =$

$\dfrac{3X+3-4+2(X^2-1)}{(X+1)(X-1)} =$

$\dfrac{3X+3-4+2X^2-2}{(X+1)(X-1)} =$

$\dfrac{2X^2+3X+3-4-2}{(X+1)(X-1)} =$

$\dfrac{2X^2+3X-3}{(X+1)(X-1)}$

(11.) $\dfrac{3a}{2a+6} - \dfrac{4a}{3a+9} =$

$\dfrac{3a}{2(a+3)} - \dfrac{4a}{3(a+3)} =$

$\dfrac{3a}{2(a+3)} \cdot \dfrac{3}{3} - \dfrac{4a}{3(a+3)} \cdot \dfrac{2}{2} =$

$\dfrac{3(3a) - 2(4a)}{(2)(3)\,(a+3)} =$

$\dfrac{9a - 8a}{6(a+3)} =$

$\dfrac{a}{6(a+3)}$

(12.) $\dfrac{5}{3x-15} + \dfrac{2}{5x-25} =$

$\dfrac{5}{3(x-5)} + \dfrac{2}{5(x-5)} =$

$\dfrac{5}{3(x-5)} \cdot \dfrac{5}{5} + \dfrac{2}{5(x-5)} \cdot \dfrac{3}{3} =$

$\dfrac{5(5) + 2(3)}{3(5)\,(x-5)} =$

$\dfrac{25 + 6}{15(x-5)} =$

$\dfrac{31}{15(x-5)}$

(13.) $\dfrac{7a}{a^2+2a-3} + \dfrac{2a}{a^2+5a+6} - \dfrac{3a}{a^2+a-2} =$

$\dfrac{7a}{(a+3)(a-1)} + \dfrac{2a}{(a+3)(a+2)} - \dfrac{3a}{(a-1)(a+2)} =$

$\dfrac{7a}{(a+3)(a-1)} \cdot \dfrac{(a+2)}{(a+2)} + \dfrac{2a}{(a+3)(a+2)} \cdot \dfrac{(a-1)}{(a-1)} - \dfrac{3a}{(a-1)(a+2)} \cdot \dfrac{(a+3)}{(a+3)}$

$\dfrac{7a(a+2)+2a(a-1)-3a(a+3)}{(a+3)(a+2)(a-1)} =$

$\dfrac{7a^2+14a+2a^2-2a-3a^2-9a}{(a+3)(a+2)(a-1)} =$

$\dfrac{6a^2+3a}{(a+3)(a+2)(a-1)} \quad \text{or} \quad \dfrac{3a(2a+1)}{(a+3)(a+2)(a-1)}$

(14.) $\dfrac{x}{x^2-9} + \dfrac{5}{x+3} - 3 =$

$\dfrac{x}{(x+3)(x-3)} + \dfrac{5}{x+3} \cdot \dfrac{(x-3)}{(x-3)} - 3 \cdot \dfrac{(x+3)(x-3)}{(x+3)(x-3)} =$

$\dfrac{x+5(x-3)-3(x+3)(x-3)}{(x+3)(x-3)} =$

$\dfrac{x+5x-15-3(x^2-9)}{(x+3)(x-3)} =$

$\dfrac{-3x^2+6x+12}{(x+3)(x-3)} =$

$\dfrac{3(-x^2+2x+4)}{(x+3)(x-3)}$

(15.) $\dfrac{a}{a-2} + \dfrac{3(2a-9)}{a^2-5a+6} =$

$\dfrac{a}{a-2} \cdot \dfrac{a-3}{a-3} + \dfrac{3(2a-9)}{(a-2)(a-3)} =$

$\dfrac{a(a-3)+3(2a-9)}{(a-2)(a-3)} =$

$\dfrac{a^2-3a+6a-27}{(a-2)(a-3)} =$

$\dfrac{a^2+3a-27}{(a-2)(a-3)}$

(16.) $\dfrac{y-2}{4y} - \dfrac{3y+5}{3y} =$

$\dfrac{y-2}{4y} \cdot \dfrac{3}{3} - \dfrac{3y+5}{3y} \cdot \dfrac{4}{4} =$

$\dfrac{3(y-2)-4(3y+5)}{12y} =$

$\dfrac{3y-6-12y-20}{12y} =$

$\dfrac{-9y-26}{12y} =$

$\dfrac{-1(9y+26)}{12y}$ or $\dfrac{-(9y+26)}{12y}$

(17.) $\dfrac{3a-b}{2a-b} + \dfrac{3a^2}{4a^2-b^2} =$

$\dfrac{3a-b}{2a-b} \cdot \dfrac{2a+b}{2a+b} + \dfrac{3a^2}{(2a+b)(2a-b)} =$

$\dfrac{(3a-b)(2a+b)+3a^2}{(2a+b)(2a-b)} =$

$\dfrac{6a^2+ab-b^2+3a^2}{(2a+b)(2a-b)} =$

$\dfrac{6a^2+3a^2+ab-b^2}{(2a+b)(2a-b)}$ or $\dfrac{9a^2+ab-b^2}{(2a+b)(2a-b)}$

(18.) $\dfrac{1}{a+b} + \dfrac{1}{b+a} =$

$\dfrac{1}{a+b} + \dfrac{1}{a+b} =$

$\dfrac{1+1}{a+b}$

$\dfrac{2}{a+b}$

(19.) $\dfrac{a-b}{a+b} + \dfrac{a+b}{a+b} =$

$\dfrac{a-b+a+b}{a+b} =$

$\dfrac{2a}{a+b}$

20. $\dfrac{6}{x^3+4x}+\dfrac{3x}{x^2+4}-\dfrac{2}{x}=$

$\dfrac{6}{x(x^2+4)}+\dfrac{3x}{x^2+4}-\dfrac{2}{x}=$

$\dfrac{6}{x(x^2+4)}+\dfrac{3x}{x^2+4}\cdot\dfrac{x}{x}-\dfrac{2}{x}\cdot\dfrac{x^2+4}{x^2+4}=$

$\dfrac{6+3x(x)-2(x^2+4)}{x(x^2+4)}=$

$\dfrac{6+3x^2-2x^2-8}{x(x^2+4)}=$

$\dfrac{x^2-2}{x(x^2+4)}$

Additional Exercises

Combine into a single fraction and simplify.

1. $\dfrac{1}{x+5} + \dfrac{2}{y}$

2. $\dfrac{1}{x} + \dfrac{3}{x+2}$

3. $\dfrac{4}{x-3} + \dfrac{1}{x}$

4. $\dfrac{3}{x+2} - \dfrac{2}{x+3}$

5. $\dfrac{x}{2x+4} - \dfrac{1}{2}$

6. $\dfrac{y}{3y+6} + \dfrac{y}{y+2}$

7. $\dfrac{2}{x+4} + \dfrac{-2}{x-4}$

8. $5 + \dfrac{1}{x}$

9. $\dfrac{1}{a} + \dfrac{1}{b} - 2$

10. $\dfrac{1}{x} - \dfrac{2}{y} + \dfrac{3}{xy}$

11. $\dfrac{y}{x^2} + \dfrac{3}{xy} - \dfrac{1}{y}$

12. $\dfrac{5}{x+5} - \dfrac{2x-3}{x^2-25}$

13. $\dfrac{2x+3}{x^2+2x+1} - \dfrac{5}{x+1}$

14. $\dfrac{2x+5}{x^2-x-12} - \dfrac{2}{x-4}$

15. $\dfrac{3}{x-6} + \dfrac{x-2}{(x-6)^2}$

16. $\dfrac{2}{3x+6} - \dfrac{x-4}{x^2-4}$

17. $\dfrac{2x + 5}{x^3 - x^2} - \dfrac{3}{x - 1}$

18. $\dfrac{5}{x^2y - xy} + \dfrac{2}{x}$

19. $x - \dfrac{x - 5}{(x - 5)^2}$

20. $\dfrac{4}{x - y} - \dfrac{-3}{-x + y}$

21. $\dfrac{4}{x - y} - \dfrac{6}{y - x}$

22. $\dfrac{2}{x + y} - \dfrac{x - 3}{y + x}$

23. $\dfrac{x + 2y}{x - y} - \dfrac{x + 2y}{y - x}$

24. $\dfrac{2x + 3}{x^2 - 9} + \dfrac{2}{3 + x} - \dfrac{4}{3 - x}$

25. $\dfrac{5}{2x} - \dfrac{3}{2x^2 - 10x} + \dfrac{1}{5 - x}$

Index

Addition of algebraic fractions, 89–99, 113–121
Algebraic fractions, 2–8, 24, 40, 51, 60–62, 70–77, 89–99, 113–121
        addition, 89–99, 113–121
        division, 74–77
        multiplication, 70–74
        subtraction, 89–99, 113–121
Canceling, 3
Common denominator, lowest, 92
Common factor, 2
Complete common factor, 2
Denominator, lowest common, 92
Difference of two squares, 38–40
Division of algebraic fractions, 74–77
Factor, 1
        common, 2
        complete common, 2
        prime, 4
Factoring, 1
FOIL multiplication, 15–20
Fractions, algebraic, 2–8, 24, 40, 51, 60–62, 70–77, 89–99, 113–121
        addition of, 89–99, 113–121
        division of, 74–77
        multiplication of, 70–74
        subtraction of, 89–99, 113–121
Fractions with opposites, 60–62
LCD, 92
Lowest common denominator, 92
Multiplication, FOIL, 15–20
Multiplication of algebraic fractions, 70–74
Opposites, 60
Prime factor, 4
Slash, 3
Square, difference of two, 38–40
Subtraction of algebraic fractions, 89–99, 113–121